Peculiar Penguins and Funny Looking Frogs

Exploring some of evolution's stranger choices

Allyson Shepard Bailey

Original illustrations by Eleanor Loughlin

ISBN 978-1-4716-6382-6

Dedicated to Sir David Attenborough, without whom this book would never have had to be written.
....

ACKNOWLEDGMENTS
A number of people have been extremely kind and generous with their time, advice, encouragement and even photographs during the writing of this book. I am especially grateful to the following authorities in their various fields, who were kind enough to recommend reading, suggest theories and/or read over my draft chapters in order to make sure I didn't say anything too ridiculous or make any obvious errors of fact.

Any remaining mistakes, omissions or impossible theories are of course entirely my own fault.

Richard Ashworth, Colour Experience manager, Society of Dyers and Colourists (Mammal Colouration)

Bruce Beehler, Senior Research Scientist, Center for Applied Biodiversity Science, Conservation International (Birds of Paradise)

Prof. Lincoln P. Brower, Research Professor of Biology, Sweet Briar College (Monarch Butterflies)

Prof. Jennifer A. Clack, Professor and Curator of Vertebrate Palaeontology, University Museum of Zoology, Cambridge (Frogs)

Prof. Ronald Douglas, Professor of Visual Science, City University London (Mammal Colouration)

Dr. Des Gilmore, Honorary Research Fellow and Adviser of Studies in Science, University of Glasgow (Sloths)

Prof. Harry W. Greene, Dept. Of Ecology and Evolutionary Biology, Cornell University (Snakes)

Dr Andrew Kitchener, Principal Curator of Vertebrates, National Museums Scotland (Cats and Dogs)

Ron Naveen, President, Oceanites, Inc.; Principal investigator, Antarctic Site Inventory (Emperor Penguins)

J. G. M. 'Hans' Thewissen, Ingalls-Brown Professor of Anatomy, Department of Anatomy and Neurobiology, Northeastern Ohio Universities College of Medicine (Whales)

TABLE OF CONTENTS

INTRODUCTION

Most of us probably think we understand the process of evolution: that living things change over time, sometimes because of mutations in their bodies, or because of some change in climate, the area they live in or the food they eat. If the animal (or plant) that has changed is more successful than those that didn't, it will pass on that change to its offspring and eventually the whole species will inherit the new features. If the change doesn't help, it won't be passed on.

We can see how this works all around us. When deer were evolving, those who could run quickly escaped predators. Long legs help you to run faster, so longer-legged males would survive to mate with longer-legged females, producing young with an even greater tendency to have long legs. So now all deer have long, thin, strong legs, specially adapted to run fast.

When we consider an adaptation like that, the idea of evolution seems to make perfect sense. But sometimes it can look as if something has gone a bit wrong. Of all the animals you would expect to find at home in a tree, the kangaroo is probably quite near the bottom of the list. They have very long hind feet, very short forelegs and a heavy, fairly stiff tail that they use to balance when they sit upright and also as a counterbalance when they hop. These traits are very useful for moving quickly to escape predators—in fact, hopping like a kangaroo is a much more energy efficient way to move than running like a deer. So kangaroos are generally very well adapted to a life grazing the open grassland of Australia. However, they lack most of the traits climbing animals usually have, like long claws or paws (or indeed tails) that can grip branches. But in fact there is a tree kangaroo. It lives in northern Australia and New Guinea. It's smaller than the red or grey kangaroos you can see grazing throughout Australia but otherwise it looks just like them. It is not, in fact, very good at tree climbing and clambers around very awkwardly. So what is it doing up there? It's looking for food. In the area where tree kangaroos evolved, there wasn't another big animal up in the trees eating leaves. If there is a source of food going spare, some animal will always try to use it, even if it isn't best equipped to deal with it. Over time, the forces of evolution will mean the descendants of the first pioneers will become better adapted. So if we could come back in a few thousand or million years, we would see that the tree kangaroos will probably have changed their shape so that they are better suited to a life in the trees. We may think a tree kangaroo looks strange, but that is only because we are seeing it en route to a better adaptation, not because nature has got something wrong.

Although we may not be aware of it, plants and animals are always "experimenting", changing their bodies or pushing their way into new environments. Of course, these experiments aren't always successful. In fact, almost all the animals that have ever evolved are now extinct. We can see this very clearly in the fossils from a formation known as the Burgess Shale, in the Rocky Mountains of British Columbia, Canada. Around 350 million years ago, this area was once covered by an ocean. Mud and sediment settled to the bottom and was later squeezed and hardened into shale. For several billion years life on Earth had consisted of only the very simplest forms, but now there was an explosion of new types and species. Along with the mud the bodies of animals that lived in the ocean at the time settled to the bottom and were covered with sediment, and so we can see some of these new experiments, preserved in the rock as fossils. Some of them we can recognise because similar animals are still alive today: jellyfish,

Roe Deer.

Notice the long slim legs for running and the long neck for browsing leaves and keeping an eye out for predators. Large ears also help warn the animal of approaching

Tree Kangaroo. (© Susan Flashman/ Shutterstock.com)

7

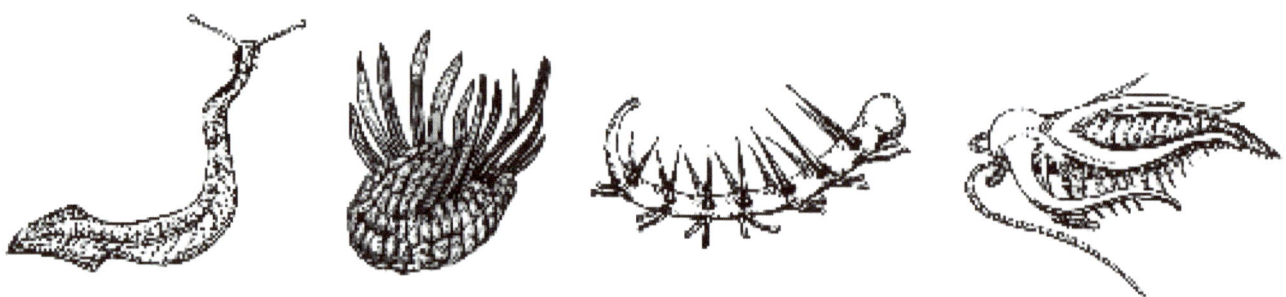

Some of the more unusual creatures from the Burgess shale. From L to R: *Pikaia, Wiwaxia, Hallucigenia* and *Marrella* (© Eleanor Loughlin)

segmented worms, molluscs like clams and snails. But some look so strange to us that they could be aliens. One appeared to have seven limbs on one side and seven tentacles on the other. With the grim humour of desperation the scientists named it "*Hallucigenia*". In point of fact, it now appears that *Hallucigenia* was originally reconstructed wrongly, and was actually a distant ancestor of today's velvet worms, but others, including one that had a body with 15 segments, a trunk in front of its mouth and five eyes, seem to have been genuinely bizarre and unlike any animals we know today. Most of these odd creatures rapidly became extinct, because in nature, just as in science, more new ideas fail than succeed. But they help us to see that nature will try almost anything.

If nature is experimenting all the time, we should be able to see it in action in the world around us. And we can, if we know where to look. Occasionally there is something in an animal's body or the way it behaves that just doesn't seem logical. But as Terry Pratchett so elegantly put it, logic is just a way of being ignorant by numbers. The point is, the apparently illogical trait must be helping the animal in some way—we just can't see how at first sight. Se we have to look harder. This book will consider some of these "kangaroos up trees" and try to understand how and why they happened. We can use information from many different places: palaeontology, genetics, chemistry, geology, even the history of our changing climate. From all these we can build up the story of these creatures and try to find out why they took the path they did.

This is neither a text book, nor the sort of book you need specialist knowledge to enjoy or understand. Specialist scientific language will only be used if it is really necessary. If you want to know more about the animals or the different branches of science we will be using, there is a list of non-specialist, easy (and interesting) to read books at the end that you might find helpful. The full scholarly bibliography is available at www.peculiarpenguins.weebly.com/read-on.html .

Having said that, it is important to understand some basic information about how scientists name and classify these things in order to give you the background against which to understand the stories we will be examining.

The earth is about 4.5 billion years old—that's 4,500,000,000. Geologists who study the planet's history have divided that time into four "eras", called Pre-Cambrian (that's the oldest), Palaeozoic, Mesozoic and Cenozoic (which includes our own time). The eras can be subdivided into periods, and some of the periods are even sub-divided again into epochs. Don't worry too much about all the names; we'll only be using a few of them. Figure 1on page 10 is a chart showing the most commonly used periods with their dates—MYA is the abbreviation for "million years ago" (rather than write it out every time, I will use this convention throughout).

As far as we know today, the very first forms of life appeared around 3 billion years ago—3,000,000,000. Our own species, human beings, only developed around 1 million years ago—1,000,000. These are huge periods of time, and it's easier to understand the timescale if we use the system developed by Sir David Attenborough in his book and television series Life on Earth. If we imagine that all the time that life has existed on earth covers one year, then each day represents about 10 million years. So the first living things, very simple, microscopic organisms like bacteria, appeared on 1 January.

Nothing new happened for quite a long time, about 8 months, in fact. The next big development was in mid-August, with the appearance of very simple plants often called blue-green algae, though in fact they aren't algae at all. (algae are very simple plants—a common sort you probably know is the green scum that forms on ponds.). Their technical name is "Cyanophytes". Simple they may be, but they changed the world forever, because when they digest their "food" (actually energy absorbed from the sun, rather than food as we know it) they give off oxygen. Up until then there was no "air" such as surrounds our planet today. Any human using a time machine to go back to the early days of life on earth would quickly choke to death. Billions of blue-greens giving off oxygen for millions of years helped to create the atmosphere we breathe now.

Gradually, the speed at which plants and animals changed, and new ones developed, began to increase. Single-celled organisms appeared in September, the first mammals early in December, and man on the evening of 31 December. We may think we are the most important animal ever to develop, but really we are newcomers. It's a bit like having your whole year's social life crammed into the last two months. The left hand side of Figure 1 shows our "year of life" and pinpoints some of the main things we will be looking at. For the most part we will only be concentrating on the last 3 months throughout this book. To help you keep track of the sequence of events, each significant date mentioned will be included in a timeline at the end of the chapter.

Just as geologists have divided and classified the history of the earth, biologists have divided and classified all the living things on it into "kingdoms". We'll really only be dealing with the animal kingdom. Within each kingdom there are five more subdivisions, each defined by more and more specific characteristics. This system helps scientists to group animals together according to important traits they share. Let's look at an animal we all know to see how it works: the tiger belongs to the **phylum** "Chordata". A phylum is the next largest grouping to the kingdom, so it has the most members, all sharing the same very general characteristic, in this case a strengthening rod running the length of their body, like our spinal column. So the tiger shares this group with all sorts of animals, from fish to frogs to man. The next grouping is the **class**, and tigers belong to "Mammalia", containing all Chordates that are warm blooded, have hair and suckle their young—which means the fish and frogs (and birds) that shared the phylum are out. Next is the **order**: "Carnivora", containing all mammals whose teeth are specialized for biting and shearing (most are meat eaters) ; the **family** "Felidae", or cats: carnivores with short skulls and well developed claws, that they can usually retract; the **genus** "Panthera", containing cats with a specially designed throat that allows them to roar (the mountain lion, for example, doesn't roar, so technically it isn't a "big" cat, but a very large small one) and the **species** "tigris". A species means a group of animals that can breed in the wild and produce offspring that will also be able to breed. (You can mate a tiger and a lion to produce either a "tigon" or a "liger", but it won't happen in the wild and neither of those new animals will be able to have cubs, so neither is a species.) Sometimes a species can be further divided into sub-species, such as the Amur tiger of Siberia and the Bengal tiger of India. These are groups of the same species that live far enough apart to have become slightly different from one another. If they stay far enough apart for long enough they may become separate species. Have a look at figure 2 on page 11 to see how each grouping fits inside the next.

All of the names I just used for the tiger are in Latin. It may seem a bit odd to use a language that no one actually speaks anymore, but in fact that is exactly the reason why scientists use it. The same animal can be found in many different parts of the world and be given different names by local people. For example, the mountain lion is also known as the painter, cougar, puma and a number of other different names. Sometimes the same name can be used for different animals: there are "robins" in Europe, America and New Zealand, and they are all very different birds. And how does an English speaker, talking about a "horse", know that it is the same as a French "cheval"?

So all scientists have agreed that each living thing will be given a name in Latin. Since no one speaks the language it won't change, and it won't give an unfair advantage to any spoken language. Usually an animal will be referred to by its genus (usually with a capital letter) and species, plus the sub species if necessary. So the Amur tiger's scientific name is *Panthera tigris altaica*. (The convention is to give the genus and species in italics). Sometimes the scientific name can tell you more about the animal, if you can work out the Latin. It can include the name of the person who first described the animal. *Equus*

9

prezwalskii, or Prezwalski's horse, is named for the Russian explorer who first discovered it. A species name that includes the word "vulgaris" means it's quite common. The American black bear is called *Ursus americanus*, while the Asiatic black bear is *Ursus thibetanus*, each name reflecting where the animal is found.

Evolution of man 31 Dec	Recent Pleistocene Pliocene	today
29 Dec	Miocene	5.3MYA
28 Dec	Oligocene	23MYA
26 Dec	Eocene	34MYA
25 Dec	Palaeocene	56MYA
Mammals become dominant Dinosaurs become extinct 17 Dec	Cretaceous	65MYA
11 Dec	Jurassic	144MYA
8 Dec	Triassic	200MYA
Mammals Birds Large reptiles(dinosaurs) 2 Dec	Permian	251MYA
Reptiles evolve 27 Nov	Carboniferous	299MYA
Amphibians evolve Fish become abundant 22 Nov	Devonian	359MYA
19 Nov	Silurian	416MYA
10 Nov	Ordovician	443MYA
6 Nov	Cambrian	488MYA
September-single celled organisms 14 August-Cyanophytes 9 Jan-bacteria like organisms 1 Jan- first signs of life	Pre-Cambrian	543MYA 4500MYA

Figure 1: Chart showing the geological periods we will be using with their approximate dates; aligned with the one year "Calendar of Life". Not to scale! The Pre-Cambrian Era should actually occupy over 90% of the chart. Some of the major events of evolution are included. MYA= "Million Years Ago"

Wherever I can, I'll use ordinary popular names—tiger rather than *Panthera tigris*. But sometimes we will have to use the Latin—for instance, when an animal is long extinct and so never had a popular name, or where we need to talk about the relationships or differences between certain groups. If it is necessary to use other scientific language it will be explained as clearly as possible.

Before we begin, just a quick note about the illustrations: all of the lovely reconstructions of extinct creatures were created by my good friend Eleanor Loughlin. They are based on the most current understanding about each animal. However, we generally don't have any information about what colour the animals were, or whether they had spots, stripes or other patterns. Rather than create something fanciful for which we don't have any evidence, we elected to make them fairly neutral colours.

Now we can start our journey. Each chapter looks at one or two questions about a certain animal or group of animals, and uses different types of evidence to try to answer the questions. Each chapter can be read separately, though I do sometimes refer back to information from earlier chapters.

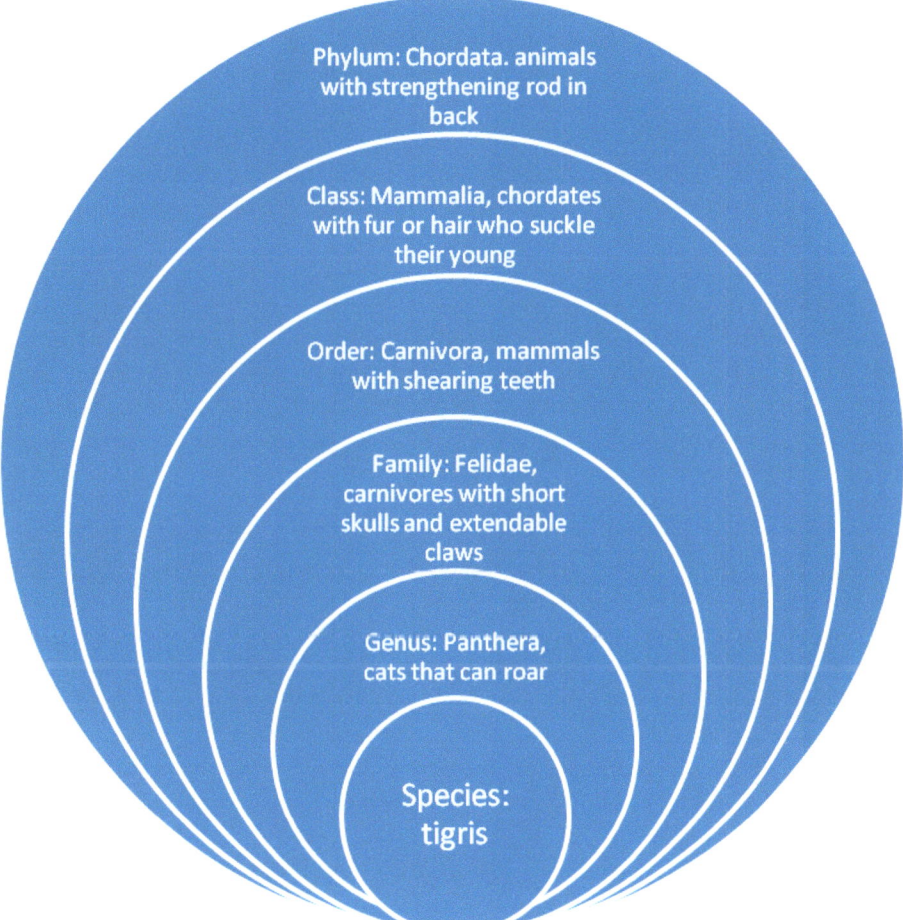

Figure 2: Classification of the tiger

CHAPTER 1: WHY WHALES?

Megazostrodon (© Eleanor Loughlin)

Of the five groups of animals with internal skeletons (known as vertebrates)-fish, amphibians, reptiles, birds and mammals- perhaps the most familiar group to us is the mammals, since that's what we are. They all share certain characteristics: they have four limbs, internal backbones, are warm blooded— which means they produce heat within their bodies rather than absorbing it directly from the sun—and are covered with fur or hair. Their babies grow inside their bodies rather than from eggs laid in a nest. Most important, they nourish their young with milk produced from a teat or breast--"mamma" is the Latin word for breast, so that is where we get the name for the group. (There is in fact a small group of mammals, the monotremes, which lay eggs but also produce milk to feed their young. They are the platypus and echidna of Australia, the only remaining representatives of the most ancient group of mammals.)The earliest mammal so far known to science is a small shrew-like creature with a long tail and pointed nose named *Megazostrodon*, which arrived around 200mya (about 12 December on our calendar), and lived quietly alongside the dinosaurs. These little animals were omnivorous, meaning they could eat anything, and were very adaptable. They survived the catastrophe that wiped out the dinosaurs and went on the flourish, spreading and changing until they occupied most corners of the earth and had evolved into numerous different species.

It's fairly easy to see the shared characteristics of mammals such as cats, dogs, horses, hedgehogs, mice and man. But with whales and dolphins family connection doesn't seem very obvious, not least because they live entirely in the water. Instead of four limbs they have two flippers, most of them have no obvious hair and they move through the water by waving their tails, like fish (though whales and dolphins move their tails up and down, fish move theirs from side to side). However, they do breathe air through lungs, rather than extracting it from the water with gills, as fish do, and their young develop inside the mother and are fed on milk.

Whales and dolphins as a group are called "Cetaceans", from the Latin word for whale. There are about 75 to 80 different species. They come in a wide variety of sizes, and have many different ways of behaving and of feeding. About three-quarters of the modern species of whales have teeth and they actively hunt fish and other marine creatures. Different species use different techniques. Sperm whales dive to amazing depths to catch squid, while around the world different groups of orca, or killer whales, use a variety of techniques to catch their different types of prey: some mob larger whales, such as greys, harassing the calves until they are weak and exhausted—for all the world like a pack of wolves cutting a caribou out of the herd. One population of orca in Patagonia actually swims up onto the beach to take sea lion pups, which is a very risky strategy: if they do not manage to get back off the beach and into the water, they will die.

The other 25% of modern whales are called mysticetes, or baleen whales. They feed by sucking in huge mouthfuls of water and straining out the tiny plants, animals and fish through horny, ragged-edged plates of baleen which hang from the roof of their mouth. There are billions and billions of these tiny organisms in certain parts of the ocean. With such a rich food source, and living in the water where they don't have to worry about carrying around a huge heavy body (the water helps support the weight far better than air, which is why you float in the water but are earth-bound on land), the baleen whales can grow to extraordinary sizes. Largest of all, the blue whale, can reach a length of over 33 metres and weigh 193 tonnes: the largest animal which has ever lived.

There are actually two things about cetaceans that are quite puzzling. First, after all the millions of years adapting to life on land, how and why did some descendants of that first tiny shrew decide to return to the sea—and then become so massive? And second, having taken to the sea so successfully, why did some whales give up hunting for filter feeding? Thanks to fossil finds from around the world, we can

(©Karina Wallton/Shutterstock.com)

(©Jo Crebbin/Shutterstock.com)

Above: Two different dental strategies. An orca, or killer whale (L) showing its sharp back curved teeth for catching large prey; and a grey whale (R) showing the plates of baleen hanging from the roof of its mouth that strain tiny food items from the water.

Below: A sense of scale. L, an orca leaps out of the water in front of two kayaks. R, a humpback "spyhopping" near a whale watching boat in Iceland. Only about one-third of the whale's full length is visible.

(© Mayskyphoto/Shutterstock.com)

(© Tatonka/Shutterstock.com)

trace most of the whales' family tree, and follow the story of the little furry mammals that slipped gradually into the water and spread throughout the oceans of the world.

Around 65mya, (December 24th on our calendar), the dinosaurs that had dominated the world for many millions of years died out. With them gone, there were all sorts of empty niches needing to be filled. All the places that dinosaurs lived, and all the foods they ate, were now waiting for another animal to use them. So those species that survived the catastrophe that destroyed the dinosaurs did what life always does when a new opportunity arises: they moved, changed and adapted to fill those niches.

It is hard to think of something as a "niche" that covers two-thirds of the earth's surface, but to these early mammals that was just what the oceans of the world were. There were many kinds of water-dwelling dinosaur that disappeared at the same time as their cousins on land, and there was food available in the water for any that could adapt to harvest it. It was the mammals that took this new opportunity, and so a few species returned to the sea their ancestors had abandoned some 350 million years before.

The extinction of the dinosaurs wasn't the only thing that helped mammals to develop. The climate then was quite different to what it is today. For one thing, there seems to have been much more oxygen in the atmosphere. Also, the climate was warmer than it is today and it was surprisingly similar all over the world: fossil species that could only survive in sub-tropical conditions today have been found in rocks that were formed in this period in Britain and even arctic Canada, places that very definitely don't enjoy sub-tropical conditions now. Temperatures on average varied by about 5°C between the equator and the poles (compared with 25° today), so the climate in Britain would have been more like that of Tenerife. The animals of this time found themselves with plenty of elbow room and a nice warm environment in which to expand. Mammals were quick to take advantage of these conditions: some 20 different orders of mammal had evolved by around 25 December. The easy climate meant that these new species could expand quickly into new areas without having to slow down and change their bodies to deal with new conditions.

Quite a lot of fossils have been found that allow us to follow the early family tree of the whale. In the most general way the story seems to go something like this: By around 45mya, mammals had developed into a number of groups. Some were meat eating predators, such as the creodonts, and others were grass eating animals with hooves—similar to today's antelope—like the condylarths. Many of the hoofed animals we see around us today can be classified by the number of toes they have. Perissodactyls have an odd number: horses have one, rhinoceroses three. Animals with an even number are called artiodactyls, and include hippos, which are the closest living relatives to whales. In the words of Richard Dawkins, a whale is what a hippo would be if it didn't have to deal with gravity on land.

Indohyus (© Eleanor Loughlin)

The hippo may be the closest living relative to cetaceans, but what is the earliest known ancestor of the group? We know that cetaceans as a group evolved from land living mammals, but which ones? Research in the last few years has indicated that a group of artiodactyls known as raoellids are probably the earliest ancestors of modern cetaceans. They evolved around 55-45 mya. Most have been found in the area of the Himalayas. The best known species is called *Indohyus*, a cat sized animal with a long snout and limbs. Each of its toes ended in a hoof. *Indohyus* also had traits that show its link to the water. The long bones of its legs have thicker walls and less spongy marrow in the centre than most other mammal bones. The same thing can be seen in hippos, and is recognised as an adaptation used by animals that spend time wading or walking along the bottom of a body of water—the thicker bone acts as ballast and keeps them steady on their feet. Chemicals preserved in *Indohyus'* teeth also show that it lived in the water, though apparently only in fresh water. However, the move to the sea had begun. In fact, the evolution of whales from now on is so fast that we need to change from using a calendar to using a diary—see figure 3.

Around 50mya, a related group of animals developed around the rivers and estuaries at the edge of the warm shallow Tethys Sea, in the area which is now Pakistan. As a group they are known as "Pakicetids", or whales from Pakistan. They varied from fox- to wolf-sized, possibly furry or possibly smooth skinned like a hippo. They had the thickened leg bones of *Indohyus* (the technical term is "osteosclerotic"), but their eyes were set close together on top of their heads, rather like a hippos. Some of the details of their skulls were quite different from the raoellids. Their surfaces of their teeth were

Pakicetid (© Eleanor Loughlin)

worn in a pattern that suggests they ate fish, while the area behind the eyes and the joint for the lower jaw were quite narrow. These differences would have affected not just the way Pakicetids chewed their food, but all the nerves and muscles in that area. So the information from their eyes, nose and mouth may have been processed in their brain quite differently as well.

At first sight there seems to be little to connect these small, possibly furry creatures with present day whales and dolphins. Their fossils have so far been found only in deposits associated with fresh water, rather than the open seas, and unlike modern cetaceans who don't chew at all, they had large chewing muscles.. They probably had limited, if any, hearing underwater, while today's whales and dolphins have excellent hearing. Toothed whales even use echolocation to home in on their prey: they make a very high pitched noise that bounces back from the objects around them. In the same way that we can hear whether a noise is coming from in front of or behind us, the cetaceans can "see" where the echoes are coming from. Baleen whales don't use echolocation, but humpbacks and other baleen whales sometimes communicate by using long, complex and sophisticated "songs".

The hyrax, one of the closest living relatives of the elephant. Although it looks superficially like a rodent, details such as the teeth and the feet are quite different. (©Allyson Shepard Bailey)

So what has led scientists to identify a connection between the Pakicetids and modern cetaceans? Well, there are many ways to prove relationships between animals that seem to have no visible connections. For example, one of the nearest living relatives of the modern elephant is the hyrax, a creature roughly the size and shape of a rabbit. The real clues can be tiny: the detailed structure of a joint, the shape of a tooth. With today's advanced technology scientists can also trace many relationships based on similarities in DNA. Evidence from other fields can be used too: different types of chemicals in a fossil bone or tooth can tell us about an animal's environment or diet. Changes in a baby animal as it develops in the womb or egg, or the development of an immature animal can sometimes tell us quite a bit about the evolution of the species. For example, baby hoatzin birds, in Guyana and Venezuela, have tiny claws on the front edge of their wings, which help them to clamber about their mangrove swamp homes. These claws, which are hidden in feathers as the bird grows, aren't found in any other modern bird, but we do see them in fossils of some of the earliest birds. Watching the baby hoatzin helps us to understand how the ancient birds might have moved around. Visually there is nothing that would make you think that *Indohyus* had any relationship to a whale, but clues such as the development of the ankle show links with artiodactyls (and later extinct cetaceans, though modern ones don't have ankles), while the ectotympanic, the bone surrounding the middle ear, is thickened on the internal side, a unique feature of cetaceans. In the case of the Pakicetids, the giveaways were the teeth and ear bones. Their strong jaws and teeth were like those of the early raoellids, but the way the teeth had worn down was different and seems to show that the Pakicetids were active hunters, like the modern toothed whales.

As it was in *Indohyus* before them, the ectotympanic, the bone surrounding the middle ear, is thickened on the internal side and their ear bones were enclosed in a shell shaped something like a grape. Both are features unique to cetaceans, though having said that, the Pakicetid ear bones were still anchored whereas the modern toothed whales' ear bones float loose in their skull, surrounded by a mass of fat. They probably had limited hearing underwater, but perhaps were developing an ability to hear vibrations through their jaw when it was resting on the ground, as turtles and tortoises can sense footsteps through their shells. The similarities of the Pakicetid ear to those of modern whales were however enough to make these the earliest creatures so far discovered to be given a name that includes "cetid" or "cetus", to indicate their relationship to whales.

Ambulocetus (© Eleanor Loughlin)

Within a few million years a new type of proto-whale had emerged, called *Ambulocetus*, the "walking whale". This stocky powerful animal grew to about the size of a large sea lion—some 2.5-3m long. It had a strong head and neck with sharp teeth. The legs were thick and splayed, with long feet—the toes ending in hooves. The tail was long but without the wide "flukes" whales today have at the ends of their tails that look rather like the blades of an oar. *Ambulocetus* probably swam rather like an otter, swishing its tail back and forth and kicking its feet. It lived in shallow marine areas and may have hunted by lying in ambush and rushing out at its prey like a furry crocodile. In at least one important way *Ambulocetus* was moving closer to modern whales: it had a special soft tissue connection between the jaw and the middle ear, rather than a bony link. So it was moving toward the floating ear bone of modern cetaceans, and developing better hearing underwater.

It is no easy matter for a land living mammal to take to the water, as anyone knows who has watched a dog retrieve a floating stick. So over the next 10 million years (only a day on our calendar) many of the features we think of as characteristic of mammals, such as four limbs and fur, were changed or lost altogether as the cetaceans became more and more committed to living in the ocean. *Pakicetus* and *Ambulocetus* had lived and hunted in fresh or at most brackish water—that is, partly fresh and partly salt, usually found near the mouths of rivers.

The next stage on the journey is represented by creatures such as the Remingtonocetidae, long snouted animals with poor vision but apparently good underwater hearing. They also show an adaptation that may have laid the basis for the wonderful acrobatic leaps and spins that some whales and dolphins can perform. The sense of balance in mammals is closely linked to the "semi-circular canals", 3 tubes at right angles to each other in the middle ear. It is the movement of fluid in these canals that helps us recognise when we are right sides up and maintain our balance—one of the reasons you get dizzy if you spin around is because the fluid in your ears is moving fast and erratically. The Remingtonocetidae had smaller semi circular canals than earlier cetaceans, which would have made their sense of balance much less sensitive and allows their descendants to perform their acrobatics without getting dizzy. The Remingtonocetidae probably lived in muddy bays, using their limbs more for propulsion in the water than for walking on land.

Protocetus (L) and *Kutchicetus* (© Eleanor Loughlin)

A bottlenose dolphin leaps out of the water. The blowhole is clearly visible on top of the head, almost directly above the eyes.
© Janprchal/ Shutterstock.com

Protocetus and *Kutchicetus* flourished around the same time as the Remingtonocetidae (some 49-43mya). They marked the real commitment of cetaceans to the sea. They were like very large sea lions or walrus. Long limbs are of no use in the water, because they don't push well against it. So the whales' forelegs, while retaining much the same arm and finger bones as our own, became short, broad paddles, while the hind limbs began to dwindle away altogether. The spine grew more flexible and the tail more powerful, with those horizontal flukes to help them move more quickly through the water—just as we use broad bladed oars to row our boats.

External ears are not much use underwater either, so they disappeared, and the skull began to "telescope", making both upper and lower jaws stick out in front of the nostrils, which began moving further up the head. The nostrils of most mammals, including ourselves, are at the front of the face, and point either down or sideways. In modern whales and dolphins they lie almost on top of the head and so are as close as possible to the surface of the water. The result is that they can now breathe by bringing just the top of the head out of the water (Hippos have a similar arrangement). They also breathe much more efficiently than we do, pushing much more used air out of their lungs with each breath.

The whole shape of the body became more streamlined to aid fast, smooth swimming. The thick fur that enabled mammals to spread to so many habitats wasn't so useful in the water. Fur may be warm, but it traps little air bubbles, making diving difficult and increasing drag. It also requires quite a lot of attention to keep it in good condition. Otters, for example, have to spend a great deal of time combing and cleaning their fur, so that they don't become chilled out of water or too water-logged when they swim. Obviously this would be very difficult for an animal with no real legs or paws, so by 40mya whales had lost most of their hair, and replaced it with a thick (up to 50cm in some cases) layer of fat, or blubber, under the skin. Fat floats better than muscle tissue and so blubber doesn't only provide warmth; it helps the animal float and swim. The thickness of blubber varies throughout the year. Like bears providing for hibernation, some whales today build up a great reserve of blubber at their winter feeding grounds, and then live off this stored fat when they head for warm water to mate and calve: warm water is a better place for new baby whales, but doesn't produce as much food as cold.

Basilosaurus (© Eleanor Loughlin)

Between 49 and 40mya, cetaceans began to spread out from the Tethys Sea. Fossils of the Protocetidae have been found all over the world. A new type of cetacean called basilosaurids flourished between 41 and 35mya. When their fossils were first discovered they were identified as reptiles (the "saur" part of the name comes from the Latin for lizard), but as more and more specimens were found, distinctive features led to the identification of them as cetaceans. Smaller basilosaurids, known as dorudontines, could still bend their flippers at the "elbow", and their nostrils were still located part way down their snout, but otherwise they were very much like modern whales. There were some 11 different species of basilosaurid, ranging in length from 4 to 16m. They had very long, flexible bodies, with relatively small heads and long snouts, and looked a little bit like popular conceptions of the Loch Ness monster.

Dorudontine (© Eleanor Loughlin)

Once the cetaceans had adapted to live in the water there was no going back. We're not going to follow their story in detail, because their development from now on was a matter of adapting to life in the seas and rivers. Various different groups and species evolved and then became extinct, until eventually the species we know today prevailed.

Why did whales take to the water in the first place? At the beginning of the chapter we said that they were expanding to fill the niche vacated by the extinction of the giant sea going reptiles. But of course this is too simple. Pakicetids didn't stand on the shores of the Tethys Sea and decide that it would be a good idea to learn to live in the ocean because there was so much food there. In fact the first cetaceans stayed in fresh water and only moved very gradually into the sea. So what made the rivers and swamps attractive?

Let's go back to *Indohyus*. In the reconstructions it looks rather like a small deer with a rat's face and the muscular tail of an otter. There's certainly nothing in its appearance to suggest it spent much time in the water. However, we need to remember those heavy leg bones—despite appearances this animal did spend enough time in water to require this adaptation. Why?

The answer may come from a modern day creature, the chevrotain or mouse deer. These tiny (generally around 35cm tall at the shoulder) deer are native to southern Asia (with one species living in Africa). Like all deer they are browsers, eating leaves or fallen fruit. Such a small animal is obviously quite vulnerable to predators, and the chevrotain has an interesting method of defence: if attacked it will jump into water and walk along the bottom until it is safe to emerge. Could *Indohyus* have used a similar strategy? Unfortunately, we don't know enough about the diet of the raoellids to be absolutely sure, but as has already been mentioned, the Pakicetids' teeth were quite different from those of their

The chevrotain, or mouse deer
(© Allyson Shepard Bailey)

predecessors. So it looks as if *Indohyus* wasn't eating fish. The changes in its leg bones that allowed it to move through the water in a stable fashion evolved before the change in diet. The very first cetacean ancestors, it seems, took to the water not for food but to escape predators.

Once there they could easily turn to finding food in their new environment. The first true cetaceans, already feeding in the swamps and estuaries, must have started sniffing around the shores of the sea and found a food source for which there was not too much competition. Slowly they followed the fish and other prey out into the open ocean. Here they found an environment with some welcoming niches ready to be filled, and within 15 million years they had become both widespread and successful. A change like this is called the "Baldwin effect": an enterprising individual discovers a new behaviour or way of feeding that is successful, and is imitated by others. This sets up a pressure within the natural selection process as those good at the new behaviour will probably be more successful at surviving and breeding, and so the trait will become accentuated over time. So those proto-whales who were best at swimming, or had longer snouts to catch fish, did better and forced their descendants down the road to the modern whales.

Having got the whales into the water, so to speak, we still have to answer the second of our questions. The early cetaceans all had full sets of teeth, most of them very big and sharp. So why did some species elect to do away with their teeth and feed using baleen instead?

Whale baleen (© Jason Mintzer/ Shutterstock.com)

We need to start by understanding when and how baleen feeding began. Baleen itself is a soft tissue and so doesn't fossilise well, so we don't have many ancient examples of it. However, like all living tissue it is nourished by blood and in the hard bones of living Mysticetes we can see the structures that allow that blood to reach the baleen. A "foramen" (plural "foramina") is a hole in the bone that allows blood vessels and nerves to pass through. Mysticetes have different foramina from toothed whales, because of the need to have a blood supply to the baleen. The grooves in the surface of the bone along which the blood vessels run (they are known as "sulci") are different too. So when a fossil skull is found with these specialised features in the bone, we can guess that the whale had some form of baleen.

Mysticetes are different from the toothed whales in a number of other ways as well. During their development mysticete foetuses do start to grow teeth. But they never erupt through the gums and are re-absorbed into the body before the baby is born. They not only lack teeth, they can't use echolocation either, though as we said before they do sing. The tiny creatures that form the diet of baleen whales don't have to be actively hunted, so there was no need for echolocation. Their eyes are also generally fairly small, but you don't need to see very well to gulp krill. Perhaps the largest, most visible difference between the two groups can be seen in the baleen whales' skull. It became much larger in proportion to the body: in some species the head is almost one-third of the total length. Baleen whales also have a loosely hinged lower jaw. Unlike ours, it is in two parts, joined by a ligament at the front (where the point of our chin is). The jaws are also bowed outward. You don't need to hunt krill and plankton, but you do need to eat an enormous amount of them, and the loose lower jaw and huge skull allow some baleen whales to open their jaws so far they can swallow millions of krill at once—a blue whale can take in over 70 tonnes of water in one feeding episode.

Although today odontocetes have teeth and mysticetes don't, the first mysticetes did in fact have a perfectly good set of teeth. *Janjucetus*, a small mysticete that lived around 27-24mya, was discovered recently in Australia. It had formidable hunting teeth and large eyes, and shows the earliest mysticetes may have been more diverse than has been realised before. There are still gaps in our knowledge, but it looks as if the process went something like this: around 37mya the mysticetes split from the odontocetes. The first mysticetes still had a full set of teeth but were distinguished by their broader "rostrum" (the "beak" or snout of the whale). Over the next few million years they grew smaller and began to develop the specialised jaws discussed above. Eventually species evolved that still had teeth but also had some baleen, such as *Aetiocetus weltoni*. They were probably filter feeding rather than hunting (The crab eater seal does the same thing today, filtering food through the cusps on its teeth). Once you start filter feeding it soon becomes apparent that teeth alone probably aren't the best mechanism to use: too much available food would slip through the gaps in such a coarse sieve. So fairly quickly (by about 32my) the mysticetes stopped using teeth at all, and instead developed full sets of baleen that would capture much more of the tiny food available in the water.

But why did baleen feeding develop at all? One answer may lie with the changes the globe was undergoing some 30 mya. The continents are always moving and changing. A map of the earth millions of years ago looked very different to what we see today. During the period those first loose-jawed whales were developing the continents were shifting and splitting. The area we know as Antarctica separated from Australia and settled over the

Aetiocetus weltoni, an early Mysticete. Notice how it has both teeth and small baleen plates (© Eleanor Loughlin)

South Pole. This changed the flow of the ocean currents and sent cold water moving northward. Cold water holds oxygen better than warm, and can support more life. These cold currents were full of tiny animals and plants. Whales moving out from their birthplace in the Tethys Sea discovered these rich spreading streams just waiting to be exploited. They could have hunted creatures such as large fish that themselves fed on the tiny fish and plankton that swarmed in these water, but that would have required a great deal of effort. By cutting out the middle man and learning to use this food source directly some whales made their lives a little bit easier. On the almost endless and easily caught diet they grew larger than ever. In fact, animals the size of modern blue or grey whales are now too large ever to return to a hunting lifestyle: it would be impossible for any hunter to successfully catch enough prey to maintain such enormous bodies, but the abundance of the tiny creatures they now eat is more than enough to keep them going. They had another incentive to grow larger as well: water rich in these nutrients tends to be found in higher latitudes, where it is cold—too cold for cetacean babies. So the mysticetes evolved a system of migration: feeding in the high, cold latitudes then moving to the warmer areas to give birth and raise their young until they are large enough to both endure the migration back and survive in the colder water. It's easier and faster for a large animal to migrate than a smaller one, so that would push the baleen whales towards increased size too.

At first glance there seems little to connect the whales with the more familiar mammals we see around us every day. But as we have seen, they are just as much a part of the family as dogs or cats. It is only their unusual lifestyle that makes them appear so different. Over the centuries man has puzzled over the whales, feared and loved them, killed and protected them. Perhaps one reason we have had such diverse feelings about them is a lack of understanding. Now we can hope that our understanding will increase as new discoveries and scientific techniques will further unravel the lives and history of these, perhaps the most mysterious of our fellow mammals.

25 (70-61mya) Extinction of dinosaurs 65mya Palaeocene	26 (60-50mya) First Raoellids	27(50-40mya) Eocene Midnight: *Pakecetus* 1:00am: *Ambulocetus* 2:00am Remingtonocetidae 4:00am *Protocetus* 6:00am *Kutchicetus* 9:00pm Basilosaurids	28 (40-30mya) Oligocene Midnight: Blubber replaces hair 7:00am Mysticetes split from odontocetes 8:00pm: mysticetes abandon teeth
29 (30-20mya) 12:00 midnight continental shifts begin to open up Antarctica 7:00am *Janjucetus*	30 (20-10mya) Miocene 10:00 early echolocation 12:00 noon earliest known baleen	31(10mya-present) Pliocene Pleistocene Recent	

Figure 3: Whale diary for the end of December

A humpback whale and her calf. Notice how the ends of her mouth reach back nearly as far as her flippers. (© melissaf84/Shutterstock.com)

A humpback feeding: it lunges out of the water with mouth agape (L), then closes its mouth (R) and uses its tongue to push the water back out of its mouth past the baleen plates, which strain out the food. You can see how the skin of the throat is pleated so it can expand, allowing the whale to take in even more water at each gulp.(both images © Jordan Tan/Shuitterstock.com)

CHAPTER 2: THE ENDURING EMPERORS

We humans seem to have a very soft spot for penguins. There are a number of reasons for this. Their upright posture and black and white colouring makes them look a bit human, like little men in evening clothes, while their awkward waddling adds a touch of the comical. They aren't majestic and powerful like the birds of prey nor exotically coloured like many tropical birds. They don't have the attractive calls of the songbirds or the sheer intimidating size of other well known flightless birds such as the ostrich or rhea. An adult penguin has little to fear from predators on land, so they meet humans with mild curiosity and little apprehension, which is a flattering change from the wariness or aggression with which we are usually greeted. However, there is more to penguins than their pleasant personalities. They are a fascinating group of birds displaying some of the most unusual and successful adaptations of any animal. This is particularly true of the emperor penguins, whose breeding strategy borders on the masochistic, and it is on them that we will be concentrating.

Drawing of a penguin by Francis Fletcher, Sir Francis Drake's chaplain, who sailed with him in the 16th century.

It took some time for this group of birds to be properly understood by science. When in the 16th century Europeans first began to encounter penguins, they seem to have regarded them with more bemusement than affection. They considered them tough, stupid and vicious. There was also a great deal of discussion about the exact nature of the penguin, with some regarding them as fish, others as birds; some decided they were a unique creature belonging halfway between the two. It's easy to see why these early explorers would be confused. Here was an unmistakably birdlike creature, with a pointed bill, webbed feet, reduced but recognisable feathers and wings. Yet it stood upright, far more so than any other bird; its wings were utterly powerless to lift it from the ground, and it spent most of its time in the water, where it swam (or flew) as swiftly and effortlessly as any fish. So it is perhaps not surprising that many opted to call them fish.

An additional factor in favour of this conclusion was the fact that those joining in the debate were largely of the Catholic faith, and searching for an acceptable alternative to meat to eat on Fridays. In the end penguins joined beavers and the world's largest rodent, the capybara, on the Friday menu, on the basis that they spent so much time in the water, and were so at home there, that it justified them being considered fish.

(©Jason Kasumovic/Shutterstock.com) (© Allyson Shepard Bailey). Beaver (L) and capybara. Even in the water neither looks very much like a fish.

Even today, when we have at least established beyond doubt that the penguin is a bird, there is much about them which is misunderstood or indeed not yet explained. If you say "penguin" to most people, they will picture a medium sized bird apparently wearing a tuxedo, standing on an ice floe. Like most such generalisations, this is more wrong than right. Penguins actually come in a wide variety of sizes, from the aptly named little penguin, only 41 cm tall and weighing just 1 kilogram, to the emperor, 1.1 metres and up to 37 kilos. They have some minor variations in colour and adornment (though all are largely black and white). They also thrive in a number of different climates and habitats.

There are around 17 different species of penguin (scientists are still debating the exact number). All live in the southern hemisphere, but their habits and habitats vary greatly. The Galapagos penguin lives, unsurprisingly, in the Galapagos islands, nearly on the equator, while the jackass penguin, named for its braying call, regularly dodges bathers on the beaches of southern Africa and nests among the suburban homes. However, it is true that most species of penguin inhabit Antarctic or sub-Antarctic regions. In fact, all species cling to areas where the ocean waters are cool--so none has been able to move beyond the range of the chilly Humboldt Current that bathes the Galapagos. Further north the waters become

Figure. 4: Map showing the main ocean currents. Notice how the circulation between Africa and South America, and between South America and Australia, doesn't reach any further than the equator.

too warm and don't produce enough food, with the result that penguins are effectively trapped in the southern hemisphere. The cold waters of these areas are very nutrient rich and support large stocks of fish for the penguins to feed on. However, in the higher latitudes, closer to the South Pole, the climate is very harsh: the summers short, the winters long, dark and bitterly cold, with temperatures commonly dropping to –60°C with driving gale force winds and frequent blizzards.

Obviously it is very difficult for any animal to raise young under these inhospitable conditions. Whales such as humpbacks and greys spend part of the year here, feeding heavily, then migrate to warmer areas to mate and calve. Some seals who favour the polar areas because of the abundance of food have drastically reduced the amount of time the young spend on shore, so that they and their parents can return to the more hospitable sea as quickly as possible: the hooded seal of the northern polar regions only feeds her baby for 4 or 5 days, the shortest nursing time of any mammal. The milk is so rich that the pup can almost double its weight in that time. After that, the mother returns to the sea and the pup must follow within a short time or die on the ice.

King penguins have extended their breeding time over two years. They generally start to nest in November (though some start later in the summer), and the eggs hatch after about seven weeks. By the end of the summer the chicks have reached their full adult size, but are still wearing their fluffy baby down. Without their sleek water resistant adult feathers they are unable to swim and so cannot feed themselves. The chicks remain on shore in crèches while the parents take to the sea, returning at regular

Some different penguin species in their various environments (not to scale): top L, Rockhopper.(© Rechitan Sorin/Shutterstock.com) Top centre, King (© Jan Martin Will/Shutterstock.com). Top R, Galapagos (©Alfie photography/Shutterstock.com). bottom L, African or jackass (© Pavel Plotnic/ashutterstock.com). Bottom R: Little (©Natalya Lysenko/Shutterstock.com).

intervals to feed the youngsters. As winter deepens and food becomes more scarce, the parents are forced to stay away for longer and longer periods. Many chicks die during this first winter. The survivors continue to be fed through the next summer, and finally fledge into their adult plumage in time to return to the sea with their parents when they are 14 to 16 months old.

The great emperors, largest of the penguins (twice the size of the kings) have developed what appears to be the most difficult, uncomfortable and dangerous strategy of them all. As the Antarctic winter begins, the emperors gather in colonies up to 200km inland, away from the sea, their only source of food, a walk—or waddle, or slide—which can take them many days to accomplish. The female lays a single egg in May, which rests on the male's feet: there is no material in this inhospitable area to build a nest, and an egg left exposed on the ice would survive for only a matter of minutes. Resting on the father's feet, covered by a flap of skin, the egg stays at the correct incubation temperature. The females then make the long hard trek back to the sea to feed. This is now an even longer journey than when they first arrived at the colony, because the winter ice sheet has been steadily expanding outwards from the edge of the land. Back at the colony, as the winter deepens, the males huddle together with their backs to the worst of the weather, constantly shuffling along so that all get a turn to stand in the warmest place: in the middle of the group and in the lea of the wind.

Emperor penguin adults with chicks: the babies rest on their parents' feet, away from the ice and kept warm under a flap of the adult's feathery skin. (© Gentoo Multimedia/ Shutterstock.com)

In the midst of all this the chicks begin to hatch. Like all newborns they are ravenously hungry, but the male penguins have not eaten themselves for many weeks, and can only offer "penguin milk", a protein rich secretion from their stomachs. Finally, the females return to the colony. Their first task is to transfer the tiny chicks from their father's feet to their mother's, without them landing on the ice. Wearing only in their first grey downy feathers, the chicks would freeze to death very quickly if left exposed. Once this transfer is accomplished the males can leave the colony in search of food. By now they are close to starvation: they have not eaten in some 18 weeks and have lost up to half their body weight. In fact, a male emperor whose mate is late returning may abandon his chick when his weight reaches the point of no return: if he does not get back to the sea to feed now, he will die of starvation. But before they can eat they must repeat the long journey across the ice to the open ocean.

After another 4 weeks the process is reversed, as the male penguins return to take over babysitting duties and feed the chicks, while their mates set off back to the sea. They will continue this pattern, swapping places every few weeks, until December. By this time, they hope, the chicks will be fledged and able to swim and feed themselves. The ice will soon be breaking up, and the emperors must take to the sea. Chicks that are not fully fledged may stay floating on an ice floe until they are ready to swim, but if the floe melts they will not survive.

This truly remarkable story seems to beg one question. Why? Why have the emperors developed what has been described by John Sparks and Tony Soper as this "aberrant and beautifully adapted breeding behaviour"? No other creature braves the Antarctic winter, especially with young. Clearly there must be an explanation. Unlike humans, animals don't do anything on a whim. If the strategy they adopt doesn't work, they will die out. So, unlikely as it seems to us, there must be some advantage to the emperors' behaviour.

In order to try to understand this, it may help if we try to trace the evolution of the penguins and find out how their journey led them to the Antarctic wastes. The origins of the penguins have intrigued scientists for many years, and many aspects have not yet been pinned down. For example, while no one would now deny that the penguin is a bird rather than a fish, there has been considerable debate over whether they are descended from ancestors who could fly, but then lost the ability, or if their ancestors never flew at all. While huge gaps remain in our knowledge, especially in the fossil record, it has become possible to work out the main lines of the penguins' story.

It is now generally accepted that birds evolved from reptiles. Perhaps the most famous pioneer was found at Solnhofen, in Germany. In this area there lies a deposit of extremely fine grained limestone perfect for lithography (a method of printing that requires a very smooth surfaced stone)—and incidentally for preserving fossils. In 1861 quarrymen found the amazingly detailed remains of a bird, complete with the delicate outlines of its feathers. But this bird had a bony jaw complete with teeth and a long bony tail: features lost in all modern birds, mostly to reduce weight and make flying easier. Clearly this was an ancestor of modern birds, and so it earned the name "*Archaeopteryx*" or ancient wing.

Archaeopteryx dates to the Jurassic period, about 140mya (or around 17 December). A few million years later, there were 40 species of birds in 20 families, and this number had greatly increased by the 24/25 December. Many of these families were aquatic. They must have been spreading out to exploit the expanding oceans, richer than ever after the water dwelling dinosaurs had died out, exactly as the cetaceans did. Being able to fly was certainly a help to this new type of animal seeking new areas to

Archaeopteryx and a close-up of its head, showing the teeth which modern birds have now lost. (© Eleanor Loughlin)

colonise. They had reached as far as Australia while *Archaeopteryx* was still flourishing.

Many of these early birds belonged to families which have since become extinct, but by about 38 mya (28 December), 26 of the 32 known modern orders of birds had developed—including penguins.

Unfortunately, the fossil record is so patchy that we don't have any information on the changes that took place between those late Cretaceous, mostly aquatic birds and the familiar flightless penguins of

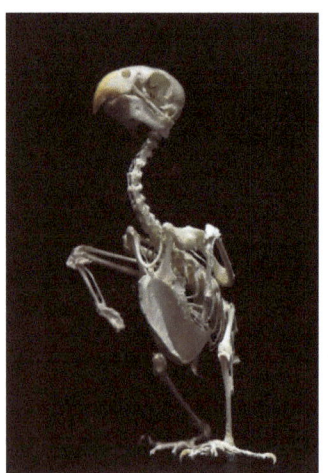

today. However, there are other clues. To begin with, almost all birds share certain characteristics which developed to help them fly. These include a sternum or breastbone with a deep "keel", that big flange of bone you see between the two halves of a chicken breast. This keel anchors the heavy muscles needed to power flight. Some modern flightless birds, known as ratites (such as the ostrich and kiwi) have developed a flattened breastbone. They no longer use their wings and so have no need of the powerful flight muscles. But penguins, which still use their wings to "fly" in the water, have retained the keeled sternum. The cerebellum is the part of the brain used for making sense of what we perceive around us, and also for controlling our muscles. In birds this part of the brain is very highly developed, because they have to make such split second decisions of speed and steering, navigation and landing. The question was once asked on the BBC's quiz programme "QI": "Why don't pigeons go to the cinema?" Despite some interesting speculation along the lines of "they don't like the popcorn", the real answer is that, because birds have such a powerful cerebellum, they can process what they see much more quickly than we can. Most films are projected at 25 frames per second. To humans, that appears to create natural motion. To a bird, it would look like a rather slow slide show. Penguins' brains have this same development, and they also have other attributes seen in flying birds. All of these features taken together indicate that they are indeed descended from flying ancestors.

The skeleton of a flying bird (© Mares/ Shutterstock.com). You can see how deep the breastbone is, compared to our own flat one

We can trace the general outlines of the penguins' evolution but the fossil record alone doesn't give us enough information to answer our questions. However, if we look closely at other evidence such as geography and climate we may be able to put together at least a reasonable story.

During the Mesozoic, from around 13 to 18 December, in the reign of the dinosaurs, the southern hemisphere was occupied by one giant landmass known as Gondwanaland. This gradually began to break up into what would become South America, Africa, Australia, India and the Antarctic. By around Dec 20th Australia and Antarctica were still connected, though the

1.8m	1.5m	3m	3m
Human	Giant Penguin	Elephant Bird	Moa

other areas had moved away. At this time the Antarctic was much warmer than it is today and heavily forested. It was in this period that the first flightless ancestors of penguins appeared.

Over the next 30my Antarctica broke free of its neighbours and drifted south, gradually growing cooler as it settled over the South Pole. At the same time, the first "proto-penguins" were developing. A modern bird which may resemble these early penguins is the diving petrel. They are fine aerial fliers but when diving they also "fly" underwater, propelling themselves with their wings, just as penguins do (rather than with their feet, like cormorants or gannets). In fact, when petrels moult they lose all their quills at once, so that for a time they are incapable of true aerial flight but can still do so underwater. They only need the flight feathers to return to the islands where they breed. By around Dec 26th the proto penguins had evolved into recognisably true penguins (known to scientists as "Spheniscidae"), flightless and fully aquatic.

We have already seen in chapter 1 how the cetaceans took advantage of the relative emptiness of the ocean after the dinosaurs disappeared. The ancestral penguins might have been still using the power of flight but they were already powerful swimmers, and must have been aiming for the same prize. As we saw in chapter 1, the gradual drifting and breaking up of the continents opened up the great southern ocean with its cold, fertile circumpolar current. By around 60mya, penguins had expanded into this rich area. There were as yet no seals or their relatives, and the cetaceans were just moving from their native area in the Tethys Sea—and in any case, most whales and dolphins eat quite a different menu from penguins. So for at least a time the penguins were the dominant "endotherms" or warm blooded creatures in the southern ocean.

There is a pattern that has repeated itself a number of times throughout the history of the earth: the most dominant and successful creatures not only diversify into many different types but grow larger. The reptiles are perhaps the most spectacular example, but it happened when insects first colonised the land, with millipedes growing to the size of cows, and much later with mammals like the mammoth. It happened to the penguins too (as well as other land living birds such as the New Zealand moas or the elephant bird of Madagascar, which grew to be over 1.8 metres tall. All of these were also completely flightless.) To date we have discovered fossils of 21 genera and 32 species of penguin which have since become extinct. Some of them grew to be 1.5m tall and weighed up to 135 kilos.

During the period 29-30 December these "giant" penguins (and others of more moderate size) occupied much of the same areas around the southern hemisphere as their modern descendants. However, the climate of these southern continents was quite different from today: don't forget how warm and consistent the world climate was while the early whales were evolving. Around 16mya world temperatures were at their highest and the Antarctic, far from being the snowy wasteland we see today, was forested and quite pleasant. Obviously the shores and islands of the continent, free of competition and with little danger from predators, all giving onto the rich circumpolar ocean, would be an ideal place for penguins to nest. Unfortunately, the change in the area's fortunes which has left much of it deeply and permanently buried in snow and ice means it is now largely impossible for palaeontologists or archaeologists to study and excavate. Only one Antarctic penguin fossil has so far been discovered, from Seymour Island. For now we can only guess whether they occupied other parts of the continent, but there is no conceivable reason why they should not have. Presumably they settled around the coasts within easy reach of their food, just as their descendants do today.

Over the next few million years several factors began to change in the penguins' world. The first seals developed, and they were both potential competitors for the penguins' food and predators on the penguins themselves. At the same time the climate began one of its periodic coolings and the ice sheets began to build up over the Antarctic. Many of the early penguin species, including the giants, gradually succumbed to these or other changes and became extinct. But quite a few of the modern genera were already in existence, and they survived to eventually give rise to the species we see today.

An emperor penguin colony. Emperors huddle together during the ferocious Antarctic storms, but even in the fairly pleasant weather shown here they are content to stay quite close together. (© Gentoo multimedia/Shutterstock.com)

Penguin fossils have been found in all the areas their modern descendants inhabit: Australia, New Zealand, South America, South Africa and of course the Antarctic. So it seems reasonable to assume that the ancestors of today's emperors had already taken up residence in or near their present breeding areas.

Now we move into the realm of pure speculation, with a few facts to hang it on. We mentioned earlier that whatever group of animals is currently dominant often tries to experiment with producing giants.

Unfortunately, these have generally not tended to survive as well as their more moderate sized relatives: larger animals require more resources and have to spend more time in conditions that are suitable for breeding to raise their young. This makes them especially vulnerable to any changes in their environment. Presumably this is why those very large penguin species died out. However, there are still two species of penguin that are notably larger than any others: the king (which can weigh up to 16 kilos) and the emperor (up to 37 kilos). Both belong to the same genus, *Aptenodytes*, and are indeed the only members of this group. Despite the fact that we tend to picture penguins standing on ice floes, as a family they are not particularly well adapted to extreme cold. One method of dealing with cold is to increase size, so it is probably not coincidental that these are the two species that spend most time in the coldest conditions.

Why would these birds choose to stay in such inhospitable conditions? As we have already seen, they are close to a particularly rich food source. Two other factors might help make this strategy attractive: a lack of competition, which we discussed above, and a lack of predators. Despite their obvious adaptations to the sea, most penguins enter it only reluctantly and with great caution. This is because they are at far greater risk in the water, mostly from seals. Few predators trouble the usually remote coastal areas penguins choose for nesting. No area is more remote or inhospitable than the Antarctic, and indeed very few emperor penguins, chick or adults, are lost to predators on land. An important factor in these calculations is that the penguins actually only required one thing from the land, a solid surface on which to lay their eggs. They had no need of food, shelter or anything else the land had to offer. So it wouldn't matter if the land was cold and barren.

Presumably the emperors did not simply take up residence in the Antarctic with conditions as they are now: the adaptations required are much too extreme to have developed quickly. So they must have begun nesting there during a period of milder climate and conditions. Like all penguins they set up rookeries in safe coastal areas. As the climate cooled further the Antarctic ice cap thickened and extended. Every time the emperors returned to nest they had to walk farther to reach land or at least "fast" ice—that is, ice that is solid year round: they could not afford to nest too close to the edge, on ice that might break up and melt before the chicks were old enough to fend for themselves.

At the same time they would have to make physical changes. The larger individuals would be more likely to survive the increasingly cold conditions. They would have more strength to negotiate the longer trek to the nesting sites. They would also be able to build up a bigger store of blubber, which would both insulate them and help them survive the lengthening fast they had to endure while awaiting their partners' return, just as a camel can use the fat stored in its hump to survive long periods in the desert. So emperors grew bigger to survive the increasingly harsh conditions of their environment—though that size would be limited to something their food source could sustain. At the same time, emperors have proportionally the smallest feet and beaks of any penguins: another device to reduce heat loss.

Most animals produce young at a point in the year when there will be an abundance of food to support their development: generally in the spring. But the Antarctic spring and summer are short, and a chick born then would probably not be mature enough to survive when the winter arrived. King penguins have solved this problem by feeding their chicks for 15 months, giving them time to become ready to fend for themselves. Kings, therefore, only breed every other year.

The emperors opted to complete their breeding in one season, but in order to raise their very large chicks so they would be able to take advantage of the spring and summer conditions they had to buy extra time from somewhere. They did it by starting their breeding during the winter. Presumably this tendency developed gradually as the birds increased in size: chicks that hatched earlier had a better chance of being ready to swim and feed when spring arrived and the ice broke up. As conditions worsened it just became harder to succeed, and it would be the bigger, tougher birds who could keep themselves and their chicks alive through the long pitch dark days in the teeth of the Antarctic winter. So natural selection would favour bigger emperors and winter-born chicks.

Another change the emperors had to make was in their behaviour towards each other. Although most penguins nest in fairly crowded colonies they are generally quite territorial and vigorously defend their nesting space and materials from any intruders. Emperors, by contrast, have no territories and indeed no nests. They have no interest in maintaining a separate "personal space". They display almost no aggressive behaviour to their neighbours. By huddling close together they greatly increase their own comfort and protection from the weather as well as those of the eggs and chicks. Indeed, within the great mass of feathers, blubber and body heat the temperature can be some 10° C warmer than the surrounding area.

We sometimes think or speak of the processes of evolution as an active choice: an organism changes to take advantage of some improved food source or habitat. For example, in chapter 1 we discussed how the first cetaceans took to the sea in pursuit of the newly unexploited food stocks there. But it seems as if the emperor penguins represent the other side of the coin. If the scenario we have developed is anywhere near the truth, it appears that the emperors found themselves a niche in which they were comfortable and reasonably successful and then, as conditions changed (and indeed worsened) they simply hung on adapting to meet the challenges of the new situation. Nature often seems determined to take the line of least resistance. The Antarctic didn't become a howling wilderness overnight, and nothing could have warned the penguins about what was coming. The changes would have occurred very gradually, and the emperors adapted just as gradually to deal with them. It was a passive response rather than an active one, but it was still successful, with the result that these remarkable birds have survived to intrigue us today.

15 Dec (170-150mya)	16 (160-150mya)	17 (150-140mya) Cretaceous Archyopteryx	18 (140-130mya)	19 (130-120mya)	20 (120-110mya) Gondwana breaking up, Australia and Antarctica still connected	21(110-100mya)
22 (100-90mya)	23 (90-80mya)	24 (80-70mya)	25(70-60mya) Palaeocene	26 (60-50mya) True penguins	27 (50-40mya) Eocene	28 (40-30mya) Oligocene Most modern bird orders including Penguins
29 (30-20mya) 12:00 midnight continental shifts begin to open up Antarctica Giant penguins	30 (20-10mya) Seals	31(10mya-present) Pliocene Pleistocene Recent				

Figure 5: Penguin timeline

CHAPTER 3: REIGNING CATS AND DOGS

If asked to name the most successful hunting animals on earth, most of us would probably answer "dogs and cats" (or perhaps "humans", but we won't go into that here). While there are other contenders for the title, these two families certainly include some of the most skilful and successful predators on the planet. From the wolves and lynx of the Arctic to the jackal and lion of the African savannas they have spread over most of the Earth and adapted to many different environments,

They are also among the most familiar predators, because of the way we have brought them into our lives and made use of their special skills. So while we may acknowledge the roles of other predators such as bears or sharks, it is cats and dogs that occupy a special place in our consciousness. We even tend to talk about them in the same breath, as if they somehow belong in a single category. Like many of our ideas about the animal kingdom, this is based on our rather self-centred human perspective. The domestic tabby keeping down mice in house or barn, the collie rounding up sheep or the husky pulling a sled: these are friends and servants we created from wild ancestors for our own purposes, so it makes sense that we lump them together in our minds.

 At first sight even the wild counterparts of these familiar pets show some remarkable similarities. Like all mammals they have four limbs, fur and a self-regulating body temperature. They bear live young which they nourish with milk. Like all mammals they are descended from that first tiny, shrew-like mammal some 200mya. Cats may have shorter faces, protractile claws (this means that the claws are normally hidden inside sheaths and can be extended when needed. They are more usually described as "retractable", which would mean that they are normally extended but can be pulled back if necessary) and more flexible, agile bodies than those of dogs. Dogs, on the other hand, tend to have an organised social hierarchy. But in general they seem to be similarly equipped, extremely successful hunters.

Yet there is one very noticeable difference between the two groups. The dogs come in a relatively small range of sizes: the huge-eared fennec fox reaches a length of only 65-70cm (all of the lengths mentioned here include the tail) and weighs about 1.5kg, while a grey wolf, largest of the wild dogs, may be as much as 1.5 or 2 m long and weigh 60kg. Cats, on the other hand, have an enormously greater range of sizes: the magnificent Bengal tiger can grow to 2.8m long and tip the scales at 300kg, while the dainty black-footed cat is no larger than an average domestic tabby: up to 75cm long and 1.6kg. This gives cats a range of lengths twice that of dogs and of weight almost 5 times as much. Nearly all of the difference is at the larger end of the scales. Most of species in both families are between 35 and 115cm long, and weigh less than 25kg (most less than 10kg). Yet for some reason a few species of big cat have outstripped these ranges by an extraordinary amount.

Why should there be such a difference in sizes in these two groups that otherwise seem so similar? They share many of the same environments and habitats. There doesn't seem to be anything inherent in their bodies to cause the difference, because in domestic animals the situation is reversed and there is a much greater range of size in dogs than in cats. There is no problem in seeing why this should be so: humans have bred the more social, easily dominated dog to the different sizes needed for different jobs from the huge hunting deer hounds to tiny toy Pomeranians. No one has ever managed to make a cat do anything it doesn't want to, so domestic cats have retained the manageable size of their most easily tamed wild ancestor (probably the North African wild cat): large enough to kill the rodents that plagued the homes and food stores of early farmers, but not so large as to be too dangerous when the fragile veneer of docility is broken (and it is just a veneer: if domestic kittens are not introduced to humans within the first week or two of birth, they will become feral and revert to wild behaviour).

If domesticated dogs are capable of achieving as wide a range of sizes as cats, why don't they do so in the wild? And, looking at it from the other end, what has enabled the wild cats to diversify in length and weight so widely? Somewhere in the evolution of the two groups we should be able to find the answer.

The creodonts we mentioned when discussing the evolution of whales in chapter 1 were the first really successful mammalian carnivores. They developed a feature that characterises many meat eaters, "carnassials" or shearing teeth. One or two pairs of upper and lower premolars and/or molars become

Above: an adult Bengal tiger (© worldswwildlifewonders/Shutterstock.com) and a European grey wolf (© Allyson Shepard Bailey). Below, a black footed cat (©Alex Sliwa. The black bulge under its chin is a radio collar) and a fennec fox (© Allyson Shepard Bailey). All four are shown roughly to scale with one another, allowing you to see the variation in size.

modified so that they mesh together like the blades of a pair of scissors, enabling the animal to slice through the tough hide and flesh of its prey.

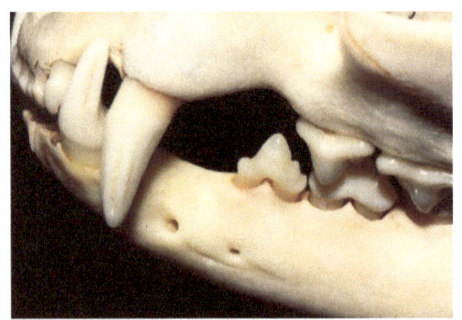

The jaw of a wild cat, showing the formidable pointed canine teeth and the scissor-like carnassials (©natursports/Shutterstock.com)

However, by around the 28th of December, the creodonts had mostly died out and their place was gradually occupied by another group of predators that developed from the same roots but had been living in their shadow: the Miacoidea. This family was characterised by small climbers, with long bodies and tails and short but flexible limbs. Although their brains were generally larger than those of the creodonts, no one is quite sure yet why the Miacoidea survived but the creodonts didn't. It has been suggested that in addition to climate change there was one major physical difference between the groups that was to blame. The creodonts, though they had the carnassial shearing teeth of carnivores, lacked molars, the grinding teeth needed to deal with other types of food. As the world grew slightly colder there was a reduction in the diversity of plants, and that may have led to a reduction in the number of prey species. However, the cooling climate meant more variation in the seasons, and so every summer and autumn there was a sudden if short-lived abundance of fruit and other plant foods. The Miacoidea had the grinding teeth that allowed them to take advantage of these foods, which may have contributed to their success. Other factors could also have helped: perhaps they had larger litters and a better survival rate. Whatever the reason, this group provided the common ancestor of both dogs and cats, although they would split apart quite soon.

With their rivals gone, the Miacoidea could expand into the now vacant niches in the environment. As so often happens when scientists try to make sense of the often fragmentary data about extinct species, there are some different theories about the road these animals took. These can be rather confusing, so let's look at them one at a time.

Miacid. (© Eleanor Loughlin)

One theory is that the Miacoidea split into two main groups: one called miacines or Canoidea, ancestors of dogs, bears, raccoons and mustelids (weasels, stoats, otters etc). The other is known as the viverravines or Feloidea, which gave rise to cats, civets, genets, mongoose and hyaenas. It has been suggested that the miacines developed in the area that became the New World, in what is now North America, and the viverravines in the Old World. Eventually, of course, members of each group crossed into the other's range across the Bering land bridge, the connection between Alaska and the eastern edge of Russia that has appeared and vanished several times throughout recent geological history as sea levels changed.

More recently it has been suggested that some very early "carnivoramorphs" (the name simply means "carnivore shaped") split into the miacidae and viverravidae. In this theory, we can ignore the viverravidae., because the miacid family included a number of different lines, one of which gave rise to the modern Carnivora. From that ancestral point the carnivores split into two groups: the Caniformia, which includes dogs, bears, racoons etc; and the Feliformia: cats, civets, hyaenas and mongoose. Both groups seem to have arisen on the supercontinent of Laurasia, the huge landmass occupying the northern hemisphere, which would eventually split into North America, Europe and Asia. (We discussed the southern one, Gondwana, when considering the development of penguins in chapter 2) If we accept the new theory about the development of Caniformia and Feliformia the original geographical separation of the two groups seems less certain. But wherever it started, the split between cats and dogs was probably established by the middle Eocene, around 42 mya or 26 December.

It seems easiest to follow the two separate, parallel lines of development in turn; and then we can compare and contrast them to see if we can find the answer to our questions. Unfortunately, this is one of the times we have to rely on scientific names, but it shouldn't get too complicated.

Caniformia: from L to R, raccoon, grey wolf, sun bear

Feliformia: from L to R, Scottish wildcat, dwarf mongoose, palm civet
(All ©Allyson Shepard Bailey). Not to scale

 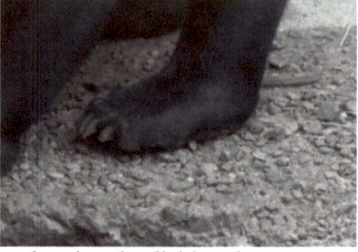

Hind leg of a Scottish wildcat, showing the digitigrades stance, and a sun bear's plantigrade foot (©Allyson Shepard Bailey)

Several features characterise a dog (or any other canid, it's easier to just say "dog"). The most obvious of them to a casual observer are the long muzzle with a large number of teeth, relatively long legs and tail and what scientists call a "digitigrade" stance. Dogs, cats and indeed many hoofed mammals stand on their tiptoes: not surprisingly, they are all creatures with a need for speed. The opposite stance, plantigrade, means to stand and walk on flat feet, with the heel in contact with the ground, a gait adopted, for example, by bears and humans. Another characteristic shared by most canid species is a fairly generalised diet, not only hunting fresh meat but scavenging and also eating insects and fruit. Animals fitting this generic dog description evolved several times after the disappearance of the creodonts. The earliest so far known, *Hesperocyon* (a Greek name meaning evening or western dog) dates to around 30mya in North America. This small (80cm long), lithe animal with short legs and long tail and muzzle probably resembled a modern mongoose more than a dog and lived the same generalised life, hunting or scavenging and supplementing its meat diet with fruit. Other dog-like mammals evolved in the same area over the next 5 million years or so and gradually spread further afield. The first to really resemble modern dogs was *Cynodesmus*, which was similar in form to the coyote. In one important respect it differed from all dogs we know: its claws were relatively thin and could be extended like a cat's. It may even have hunted like a cat, stalking and then ambushing its prey.

Hesperocyon (top) and *Cynodesmus*. Not to scale (© Eleanor Loughlin)

The *Hesperocyon* line was relatively successful for a few million years but eventually became extinct. They were, for a time, replaced by a group which had co-existed with them throughout a good part of their history. The miacine ancestors of the dogs eventually gave rise to three other groups: bears, mustelids and the procyonids, a small family of mostly tree-dwelling animals such as the raccoon and coati. On the way to producing these modern groups, there was another split in the family that resulted in two distinct groups: the bears, and the "amphicyonids" or half-dogs. They were in fact not true dogs at all, but rather had split from the same ancestor that also produced the modern bears. Unlike the earlier proto-dogs, the amphicyonids first evolved in Eurasia, and from there spread throughout the northern hemisphere. They developed into a wide variety of forms: *Amphicyon* was large and bear-like, with a largely vegetarian diet, while *Cynodictus* was smaller and more dog like. Some species resembled cats more than dogs or bears.

Amphicyon (© Eleanor Loughlin)

Eventually, by about 8mya or 29 December, the amphicyonids too became extinct. The reasons for this are not completely understood, but one theory is that they weren't very well suited to the changing climate of the time. Temperatures were starting to fall and forest gave way to grassland, where the lack of cover meant speed became more important to both predators and prey. Perhaps the carnivorous amphicyonids' normal prey died out and they could not adapt to hunting the new types, while the more vegetarian ones could not sustain themselves as the plants changed.

In any case, by this time the true dogs, genus *Canis*, had begun to diversify and spread out from their first home in North America. Of the species living today, the first two to split from the original line were the grey fox, which lives in the southern and western parts of North America and the raccoon dog of central and Eastern Europe and East Asia. The former is a handsome long-bodied animal with a brindled grey coat and white belly, tinged with red on neck, flanks and legs. This oldest of the canids retains some of the arboreal habits of its early ancestors, which most other dogs have lost—though the common European fox can climb if necessary. The grey fox, however, is quite at home in the trees, climbing with remarkable, almost cat-like agility and running along branches using its bushy tail as a counterbalance.

Grey fox. (© Leroy Gunderson/ Shutterstock.com)

The raccoon dog has the rotund shape and black "mask" of its namesake, and also climbs easily. Uniquely among dogs, it hibernates through the winter. Both it and the grey fox enjoy the widely varied diet that seems to have helped make the dogs so successful.

Raccoon dog (© Marc Herrmann/ Shutterstock.com)

At some point after these two species evolved the dog family separated into two distinct groups: the "vulpine" or fox-like and the "lupine", or wolf-like, such as the wolves, jackals and coyotes. Both types have spread all over the world with the exception of the Antarctic, Australia (the dingo was introduced into Australia by humans and is descended from domestic ancestors) and various islands. Many species live fairly solitary lives, either alone or in pairs, while others live and hunt in packs of varying sizes. The fearsome African hunting dog is the most social, living in groups of 30 or more. It may not be a coincidence that this is also the only type of wild dog to have much variation in coat colour: they bear patches and swirls of black, white, yellow and grey which must help make each individual in the large group instantly recognisable. All dogs eat a wide range of foods, but the type of diet varies greatly between species: the fennec fox is largely insectivorous, the long legged maned wolf pounces on small mammals and insects hidden in the long grass of the pampas, as well as eating various fruits. The corgi-like bush dog of South America, the Asian dhole and the grey wolf all live and hunt in packs which allow them to bring down much larger prey than they could otherwise handle as individuals, and so have a far more meat based diet than some of their cousins.

Maned wolf (©Allyson Shepard Bailey).

Dogs hunt mostly by scent, though their senses of sight and hearing are also very keen. Smell is also very important in their social lives, being used to identify individuals (and their position in the social hierarchy) and to mark territorial boundaries. Their social organisation can also help their hunting, as remarked above: no lone wolf will ever grow to the size or strength of a lion, but a pack working together can run down prey much larger than any one individual could tackle, pursuing them at that tireless lope that eventually exhausts the victim enough for the pack to pull it down. Again, co-operation tells here: the wolves' long muzzles mean their bite is much weaker than that

Painted hunting dogs (©Allyson Shepard Bailey).

of a similar sized cat, but enough jaws working together can make the kill as effectively if less elegantly.

Dogs are generalists. A basically successful anatomy has evolved to adapt it to many different habitats and ecologies. Anyone who has seen a fox raiding urban dustbins can see just how successful their behavioural flexibility can be. But in opting for a lifestyle that combines, wherever possible, speed and strength, sociability and solitude, meat eating and fruit eating, the dogs have, in a way, lost the chance to really hone themselves into outstanding representatives of any one trait.

Wolf pack (© Ewan Chesser/ Shutterstock.com)

From L to R Serval cat, dwarf mongoose and palm civet (© Allyson Shepard Bailey)

At about the same time the first ancestors of the dogs were evolving (around 40mya), the cats were beginning their evolutionary journey from the early viverravines, the family that also produced the civets, genets and mongoose. Even today it is easy to see the relationship between these groups, with their long, slender, flexible bodies and tails and quick agile motion.

Running cheetah (© photobar/ Shutterstock.com)

All cats share certain characteristics, whatever their size, habitat or lifestyle. Like dogs, they have the tip-toe stance that aids speed. However, unlike dogs their claws are normally retracted inside individual fleshy sheaths at the end of each toe, and the cat actually has to make a muscular effort to extend them. This keeps the claws needle sharp (as any cat owner can testify), excellent for grappling and tearing prey. A cat's spine is extremely flexible. The cheetah, the fastest land mammal, can bend its back far enough forward to bring its hind feet in front of its forefeet at the beginning of a stride, allowing it to effectively lengthen its stride well beyond its body length. A cheetah at full speed—which can reach 100kph—can have a stride of some 7-8m.

In fact, the cats' whole structure is very flexible, allowing them to climb, jump and twist with acrobatic grace (and, famously, always land on their feet. "Always" is an exaggeration, but if the drop is long enough a cat should be able to twist itself into an upright position before reaching the ground.)The degree of flex varies greatly: the powerful and agile leopard, though it frequently climbs trees to rest and store prey out of reach of scavengers, has to come down carefully and rather awkwardly, while the smaller, more truly arboreal margay has ankle joints which can twist almost through 180° allowing it to run down a tree almost as easily as it goes up.

Cats have a relatively short, round skull, with strong muscles powering their bite. The short muzzle means they have fewer teeth than dogs, but there is much more power concentrated in the length of those jaws, and a cat has a much stronger bite than a dog of equivalent size. Their canine teeth are

Proailurus, the ancestor of both modern cats and the extinct sabre toothed cats (© Eleanor Loughlin)

particularly long and sharp, and in many species the teeth are spaced to allow the killing bite to penetrate the spine of their prey: the spacing varies depending on what the main prey of that type of cat is. The carnassials are very efficient shears, but cats lack the grinding molars of herbivores and omnivores, including man (and indeed dogs). So cats, although extremely deadly killing machines, are rather inelegant eating machines, tearing off and swallowing small chunks of meat without really chewing.

Between 35 and 25mya (during 27 December) the early cat like creatures went through a number of transformations, from the Aeluroidea to *Proailurus*: a generalised cat-like branch of the family ("Aelurus" is the Greek word for cat). *Proailurus* may have been the ancestor of both the Felidae, modern cats, and the famous sabre-toothed cats such as *Megantereon* in Africa, Eurasia and North America and *Smilodon* in the New World. *Smilodon* carcasses form part of the remains found in the famous La Brea tar pits of

Los Angeles' Hollywood. Presumably attracted to prey animals caught in the sticky bogs, the cats in their turn became stuck and died there.

Though the sabre tooths are clearly cats, they show some important differences from the modern species. Fairly large but more heavily built than most modern cats, they were most successful in the age of large slow prey such as woolly mammoth. The huge canine teeth that give them such a formidable appearance were actually quite fragile and could have been easily snapped by a small sideways force. They were probably used to give a shearing bite to the soft tissue of the throat or belly after the prey had been subdued, not to dislocate the neck vertebrae as modern cats do.

The sabre tooth cats developed alongside modern varieties of "neofelid" (or new cats) but gradually the climate changed, the grasslands spread and the huge slow moving herbivores died out, to be replaced with the lighter, faster hoofed mammals that we are familiar with today. The arrival of humans may also have contributed to the extinction of the giants. Unable to keep up with the new prey species, *Smilodon* and its cousins also lost the struggle and became extinct.

Smilodon (© Eleanor Loughlin)

The modern cats followed a slightly different path. *Proailurus*, the ancestor of all modern cats, still had many features more reminiscent of the civets or genets, which have longer snouts and more teeth than cats. By 20mya *Pseudaelurus* had evolved: a rather generalised cat from which all modern cats are descended. This diversified into several different types, which began moving from their area of origin in Eurasia into Africa and North America around 9 mya (29 December)

These early cats were about the same size as today's ocelot, but fairly rapidly they began to evolve in to both larger and smaller forms. The first branch to split off, around 6.4mya, gave rise to all the big, roaring cats. Over the next couple of million years ancestral species of all the types we know today began to develop. There was also a further wave of migrations, as sea levels fell between 4 and 1mya. One of the early cheetahs was a giant, weighing about 120kg (twice the weight of a modern cheetah); the lynx, best known today from the northern regions of Europe and Canada, developed in Africa; the jaguars found in South America

Pseudaelurus.(© Eleanor Loughlin)

today evolved in Eurasia from an ancestor they share with leopards and lions. Small cats were evolving too, though there is less fossil evidence for them: their small fragile bones don't survive so well, particularly in the wooded environments they favoured. Some species we do know of include leopard cats, Martelli's wild cat—ancestor of the European wild cat—pampas cats and margays.

For the most part cats are solitary hunters: only lions regularly hunt in groups, though siblings of other species may join forces, and of course mothers teach their cubs. Their sense of smell is not nearly as keen as dogs' and they locate their prey by sight, stalking it with infinite patience and in complete silence until close enough to ambush it. Even the speedy cheetah will get as close as it possibly can to its prey before starting its attack: it knows it can't sustain its speed for very long. Small prey animals are dispatched with a bite through the spinal cord, larger ones suffocated with the cat's jaws clamped over windpipe or muzzle. The original medium-sized cat design seems best suited to tackling rodents and other small animals, and it seems likely that this was their original mainstay. Indeed, most cats are still fascinated by burrows that

A leopard stalking prey. It keeps its body as low to the ground as possible, its eyes fixed on the target. (© Ecoprint/Shutterstock.com)

might hold such prey, or even by something that looks like a burrow, as any domestic cat owner has seen who has left a shopping bag on the floor.

Cats are among the most specialised hunters on Earth, unlike the dogs who opted for a flexible lifestyle. Evolution has fine-tuned every aspect of the cats' anatomy to make them the finest hunters they can be: exceptionally keen eyesight, razor-sharp teeth powered by strong jaws, flexible agile bodies well camouflaged in their chosen environment and soft silent feet equipped with deadly grappling claws.

There is a down side to such narrow specialisation, however. An animal that relies entirely on hunting needs a larger territory to sustain itself than does one with a more varied diet. Even without the pressures of deliberate hunting for fur or other products (such as the tiger bone used in traditional Chinese medicine), the steady reduction in usable habitats has made life difficult for all cats. Few are now considered common, many are vulnerable and several are extremely endangered. In contrast, though many dogs have suffered from human persecution and habitat loss, quite a few species are still common and fewer are vulnerable or endangered.

Now that we have explored the evolution, development and main characteristics of the two families, are we any closer to answering our original question? Why have the cats been able to grow so much larger than the dogs? The answer must lie in the most fundamental difference between them, their diet. A bigger hunter can pull down bigger prey, which is a much more efficient way to obtain energy. A lion that has caught a zebra can relax for several days, but an insectivore or even a hunter of small rodents must work harder to collect the same number of calories. However, even a very large wolf would lack the grappling claws and strong bite needed to dispatch a large prey animal quickly and so would risk serious injury if it attempted to tackle one. A pack of very large dogs could indeed deal with prey of that size, but would need to catch a lot more to keep all their members well fed.

It would appear that the cats' dedication to hunting is responsible for their ability to grow so much larger than dogs. If a leopard spends half an hour stalking a rabbit, the result is a snack. The same half hour spent stalking an impala results in a hearty meal that can feed the cat for quite some time. Another possibility may lie in the different hunting techniques the two groups use. Those dogs that hunt are "pursuit predators"—they actively chase down their prey. You can't get up much speed if you're too big. Cats, on the other hand, ambush their prey, and size is not a problem there. So up to the point where you become too bulky, slow or noisy to hunt successfully, it seems that increased size is one option the true hunter has that the generalist lacks.

22 Dec (100-90mya)	23 (90-80mya)	24 (80-70mya)	25(70-60mya) Palaeocene	26 (60-50mya) Miacoidea	27 (50-40mya) Eocene Cat and dog ancestors split (42mya)	28 (40-30mya) Oligocene *Hesperocyon* (30mya) First cats (40mya) Aeluroidea (35mya)
29 (30-20mya) *Cynodesmus* (25mya) Amphicyonids *Proailurus* (25mya)	30 (20-10mya) Sabre tooth cats. Pseudailurus (20mya)	31(10mya-present) Pliocene Pleistocene Recent Cats spread from Eurasia (9mya) Dogs spread from North America (8mya) Modern big cats (6.4mya)				

Figure 6: Cat and dog evolution timeline

CHAPTER 4: MIGRATING MONARCHS

Monarch butterfly (© Jason Patrick Ross/Shutterstock.com)

The North American monarch butterfly is a familiar companion to many American and Canadian childhoods. Boldly and handsomely marked in black, white and orange, they can be seen throughout the summer in much of the continental United States and southern Canada, near fields or waste ground, or anywhere their main food plant, milkweed, grows.

Few people, seeing a monarch flitting across their garden, would notice anything unusual about it. In fact, the North American monarch (the scientific name is *Danaus plexippus*) has one of the most remarkable life cycles in the insect world. In common with many other species, several generations will be born each summer. The adult butterflies live only a few weeks, just long enough to mate and produce up to 400 eggs, always deposited one at a time on milkweed plants.

Milkweed shedding its seeds
(© KellyNelson/ Shutterstock.com)

This tough pale green plant exudes a gluey poisonous white sap that gives the family its name. In autumn each of the big oval seed pods release hundreds of little seeds with fluffy white thistledown-like parachutes. Monarch caterpillars feed on the leaves, somehow storing the toxins in their own bodies so that they will be unpalatable to predators. The adult butterflies retain this toxicity and the bold colouration is probably an advertisement to warn birds and other predators not to eat the poisonous mouthful. Other non-toxic butterflies have very similar markings, and lepidopterists (experts on butterflies and moths) have theorised that this mimicry has evolved to take advantage of the monarchs' protective chemistry: presumably birds seeing an almost identically coloured butterfly will assume that it, too, tastes bad.

Viceroy (© Sari O'Neal/Shutterstock.com), Monarch (©Elizabeth Spencer/Shutterstock.com) and Queen (© Richard Fitzer/Shutterstock.com) butterflies showing how species mimic each other. Not to scale

However, the last generation of monarchs to be produced, as summer begins to turn into autumn, is different. Their bodies may be slightly smaller than average, and for the most part they won't mate or lay eggs. This is known as a state of "diapause", when normal sexual processes are suspended. Barring accidents, they will live for eight or nine months. These monarchs must make an epic journey south, away from the harsh northern winter, and then return the following spring to produce the next year's first generation.

Responding to cues such as the change in temperature, the shortening days and the dying back of the milkweed, the summer's last generation of monarchs begins to fly purposefully south and southwest, gradually coalescing into large groups that stream through parts of Texas, sometimes millions strong. They all head for a small area in the mountains of central Mexico: the Oyamel fir forest—a place, incidentally, that none of them has ever seen before. They will fly on average 71 kilometres a day, though they can cover up to 320. (Populations west of the Rocky Mountains make a similar move southwesterly to the coast of California, but for simplicity's sake we will ignore them here.)

Figure 7: Map of North America showing the monarch's migration route and the general area of the Oyamel forest.

The butterflies congregate in a small area around three kilometres above sea level, on the steep, south-west facing slopes. The location must be carefully chosen. The temperature must stay in a range between 15° and 0°C, warm enough to stop them freezing, but cool enough to keep their bodies operating slowly and so reduce the loss of energy and body fat—the butterflies can't afford to use up their reserves before there are any nectar bearing plants available to feed on. The surrounding trees buffer the crowds of butterflies from the wind, rain and snow. This protection from the elements is important: a wet butterfly will freeze to death at -3°C, but a dry one can survive in temperatures that average as low as -8°.

The butterflies hang from the undersides of branches and trunks of trees, clustered together in large groups and hanging onto the needles that cover the branches. Occasionally they may shiver to warm up, or take short flights to drink. But for the most part they are motionless, conserving their strength until conditions become more favourable.

During February, as the days become longer and warmer, the monarchs start to become more active. They begin to mate. When spring is far enough advanced they will begin to fly north again. Few if any will live to reach the far north, the areas they left in the autumn. They will reach the southern states as the milkweed is re-emerging and lay their eggs. They will die here, but their offspring will emerge as caterpillars, pupate, and continue the journey north as adult butterflies. It may take several generations to reach the northern limit of their range, in full summer. Then the cycle begins again.

We are impressed by the well known migrations of larger animals: the wildebeest following the rains back and forth through Africa; the humpback whales moving from their warm breeding grounds near the equator to the cold productive northern waters; the Arctic terns' flight from one polar region to the other to take advantage of the long days and short lived glut of food in the Arctic and Antarctic summers. But the journey of the monarchs surpasses all these. The tiny creatures, 3cm long and weighing around 1g, travel

Monarchs overwintering in Mexico (© Charles Shapiro/Shutterstock.com)

up to 4000km. Other butterflies, such as painted ladies, do sometimes migrate in response to population pressures. Others hibernate, passing the cold winter months in a dormant state. But none migrate the distances monarchs do, or then go on to survive for so many months.

So how and why did this remarkable behaviour develop? Obviously any animal living in the temperate regions needs a strategy to survive the long cold winters, with all the associated problems of bad weather, lack of light and scarcity of food. Even large vertebrates have problems, and many of them hibernate or migrate to warmer areas. Tiny, short-lived, cold blooded insects must find the situation even more challenging. Many butterflies over-winter either as eggs or pupae, which are slightly less vulnerable than the caterpillar or adult butterfly stages. What makes monarchs different?

		Recent Pleistocene Pliocene	today
31 Dec			
28 Dec		Miocene	5.3MYA
27 Dec		Oligocene	23MYA
25 Dec		Eocene	34MYA
24 Dec		Paleocene	56MYA
17 Dec Flowering plants	Fleas; Praying mantis; Termites; **Butterflies**	Cretaceous	65MYA
11 Dec	**Primitive moths** Cockroaches	Jurassic	144MYA
8 Dec	Ants, bees, wasps; Caddisflies; Stick insects; Dragonflies, damselflies	Triassic	200MYA
2 Dec	Flies and mosquitoes; Lacewings; Beetles; Thrips; Stoneflies	Permian	251MYA
27 Nov	"Bugs": aphids, cicadas etc; Grasshoppers and crickets Mayflies etc	Carboniferous	299MYA
22 Nov Conifers and cycads Mosses, liverworts etc Ferns	Wingless insects: bristletails etc	Devonian	359MYA
Land Plants 19 Nov		Silurian	416MYA
10 Nov		Ordovician	443MYA
6 Nov		Cambrian	488MYA

Figure 8: The evolution of some of the more familiar insect orders, and of different types of land plants. The earliest moths appear to pre-date flowering plants and so could not have been nectar feeders. However, even today the most primitive moths chew pollen or fern spores, which is presumably what their early ancestors did

Let's start at the beginning. Butterflies and moths are known collectively as "Lepidoptera" (The name comes from two Latin words and means "scale wings"—the colours and patterns on their wings are formed from tiny overlapping scales). Although it appears that butterflies actually evolved from moths, it can be very difficult to distinguish between the two, and sometimes the distinction is purely arbitrary. A pale coloured Lepidoptera that flies at night is almost certainly a moth, and one with clubs or knobs on the ends of its antennae is bound to be a butterfly—beyond that the layman is unlikely to be able to determine which category an individual belongs to. Such tiny frail bodies are unlikely to fossilize well, so relatively little is known about the evolution of the order. They are probably the last of the insect orders to have arisen, long after the other familiar inhabitants of our homes and gardens, such as beetles, bees and wasps or grasshoppers. The earliest known fossil Lepidoptera dates from the lower Jurassic, around 144mya or around 12 December. Flowering plants, known as angiosperms, had already evolved and radiated around the world by the early Cretaceous. Lepidoptera, despite their late development, also radiated quickly, following the flowering plants that became their food source as they colonised the earth.

Some of the more spectacular butterflies and moths: L to R:green swallowtail, *Agraulis vanilla* (©Sari O'Neal/Shutterstock.com), European peacock (©FotoVeto/Shutterstock.com), Giant silkworm moth (© Reinholt Leitner/Shutterstock.com) and Giant Luna Moth (© Stephen Russell Smith photos/Shutterstock..com).
Not to scale

Today, there are over 165,000 species, flying by day or night, varying in size from tiny micro-moths no more than 15mm long to the great Atlas moth which can have a wing span of 30 cm. Most noticeable to us is the fantastic variety of colours and patterns: some night flying moths are a ghostly white or grey, but many butterflies and moths flaunt brilliant, iridescent blues and green, vivid orange, or velvety black. Some butterflies sport long swallow tails to their wings, while some moths have feathery fringed antennae.

All Lepidoptera follow the same life cycle: tiny caterpillars hatch from eggs laid by the adult female. As soon as they hatch they begin to eat voraciously, stripping leaves with a rapidity that is the despair of many gardeners. Like all invertebrates—animals without internal skeletons to support their bodies—they can't increase their size very much before they must spilt and discard their skins. Then they continue to eat and expand into the new skin. Each of these stages is called an "instar". Most species will go through four or five instars, eventually increasing in size up to 3000 times. Interestingly, the caterpillars head doesn't greatly increase in size—so by the time it reaches the last instar the head will seem much more in proportion to the body than it does when it first hatches.

Finally, when the caterpillar has accumulated enough body material to produce the adult form, it stops eating and begins to "pupate". Burrowing into the ground, or attaching itself dangling from a twig or leaf, it creates a pupa, or protective casing, from its own body. Some moths spin cocoons—the "silk worm" from which we get silk fibres is actually the caterpillar of a silk moth. Butterflies create a hard, often shiny chrysalis. Within the

Composite picture of a monarch caterpillar hatching and then eating its eggshell (L © Cathy Keifer/Shutterstock.com)and a mature monarch caterpillar. (© James A. Kost/Shutterstock.com). Not to scale

pupa an astonishing transformation takes place: the caterpillar's body breaks down and is re-assembled as the adult butterfly or moth. When the new form is complete it emerges from the pupa and flies away. No longer eating leaves, instead it sucks up flower nectar (in some species, rotting fruit, moisture from dung or other foods) through a very long, tube like tongue called a proboscis. In many species the proboscis is visible, coiled like a watch spring below the butterfly's head. The sugary, high energy nectar fuels the adult's search for a mate. The female then lays her eggs on the appropriate food plant that the caterpillars will eat. Some species will mate and produce eggs several times, but few live more than a few weeks or a month. The great Atlas moth has no mouth parts at all and so can't feed. They only live for 7-10 days, just long enough to mate and produce the next generation.

Composite photograph of a monarch butterfly emerging from its chrysalis. (© Cathy Keifer/Shutterstock.com) The whole process will take several hours

In tropical or semi-tropical areas the cycle can go on more or less continuously. The climate varies so little throughout the year that there is a constant supply of both food plants and nectar bearing flowers. As you move into more temperate regions, however, the cold winters create an increasing challenge: not only do the sources of food tend to die back, but the colder, often freezing temperatures are an extra hazard. Like all animals except birds and mammals, butterflies and moths are "cold blooded." The term is a bit misleading: their blood is not necessarily cooler than ours. It is more accurate to call them "ectothermic", from two Greek words meaning "outside" and "heat". They don't produce heat from inside their bodies as we do (we are "endothermic") but absorb it from their surroundings. That is why lizards and snakes are sluggish in the early morning, until they have absorbed enough heat to fire up their engines. This need to get heat from the environment is why there are no amphibians or reptiles in polar regions, and only a few very specialized invertebrates, that actually produce a sort of natural anti-freeze within their bodies.

So if the temperate regions are so much more difficult for butterflies than those closer to the equator, why do they live there at all? The answer, of course, is food. As the flowering plants moved out over the earth, the animals that fed on them naturally followed, evolving and adapting as necessary. A butterfly that "learned" how to deal with a slightly more rigorous climate than its fellows would be able to take advantage of food sources that others could not access.

It's important to remember that overall the earth was much warmer from the time the Lepidoptera developed until less than 40 mya when it began the cooling it is still undergoing. During the Eocene (53-34mya, about 26 December) the evolving flowering plants and the butterflies that depended on them would have enjoyed the same freedom of movement over a world with very little variation in temperature—we noticed this when discussing whales in chapter 1. In fact, the modern types of Lepidoptera we know today mostly emerged during the Miocene period, when the climate was quite cool. The open grasslands were spreading and more cold tolerant, deciduous plants (those that shed their leaves and shut down during the winter) developed that could survive the seasonal extremes. Butterflies and moths developed alongside them.

As the climate started to cool, the butterflies and moths that had colonised areas now becoming more seasonal had two choices: move back into areas with more consistently warm weather, or adapt to the new conditions. The warmer areas were presumably already occupied, there would be too much competition for food—and probably too much danger from unfamiliar predators. A new adaptation was almost certainly a better bet.

The greatest part of this adaptation would be surviving the winter, in one form or another. No food plants would be available for either caterpillars or adult forms, so they would have to ensure that they were very well fed, and then ensure they had a safe place to over-winter, safe from the coldest temperatures as well as from any predator searching for a tasty morsel to help get through the hard times. Eggs and pupae at least don't have to worry about food. All temperate zone butterflies and moths have managed to adapt one of their four stages to over-winter locally—except the North American monarch.

There are some 157 species of milkweed butterfly. All are tropical or sub-tropical, generally inhabiting forest or woodland. None of them has to deal with the problem of surviving a winter season with no food, with the exception of the North American species, the only one to have moved far into the temperate range.

Animals move into a new environment because of some pressure or threat that impacts on their food sources or ability to breed. Nectar bearing flowers, the food of the adult monarch, are common throughout many areas, and are exploited by many different butterflies and moths, so it is unlikely that any change in these types of plants spurred on the monarchs' move. Milkweed, on the other hand, is the monarch caterpillars' only food, and therefore their migration would be inextricably tied to the plant's spread.

The places milkweed could colonise would in turn be linked to the prevailing climate. It appears that the butterflies of the monarch's Danaus genus moved from South America into Central America about 3mya, towards the end of the Pliocene. The mixed grasslands of the great plains were expanding at the same time, so the butterflies could follow it north—but they would still have to retreat south during the winter to avoid the freezing temperatures. During the Pleistocene the ice advanced and retreated several times. 18,000 years ago, during the last of the Ice Ages, permafrost covered North America as far south as Chicago. "Boreal" or northern type coniferous forest stretched over much of the southern part of the continent. This was no place for temperate grassland plants.

Figure 9: Map showing the furthest extent of glaciations in North America

By 10,000 years ago the climate had warmed, the ice retreated and much of North America became covered with prairie grassland and temperate woodland: a rich and varied environment that included, of course, milkweed.

As the milkweed plants spread north the monarchs followed them. Obviously all species of butterfly using the same food plants could have "chosen" to do this, but apparently they didn't, or perhaps some did but were unsuccessful. Perhaps *Danaus plexippus* had some characteristic that made it easier for them to change their environment. Certainly today they are better adapted than most temperate zone butterflies to cold temperatures: their bodies are dark, which means they absorb heat, and larger than average, which helps them retain it. In general they are larger than most tropical species of monarch.

So some monarchs began to expand their range, pushing the edge of their territory in a search for areas where there was less competition for food. Larger individuals could fly further and endure colder temperatures. To begin with, the distances travelled would not have been very great, and as the season turned in the north they could move back. Even when the distance became too great for a single generation—remember, an adult monarch normally only lives a few weeks—they could lay their eggs, the resulting caterpillars would eat the milkweed, pupate and continue the southward journey as adults.

Gradually, however, as the climate continued to moderate, milkweed colonised areas thousands of miles from the butterflies' homeland. The plant is an herbaceous perennial, which means that it dies back each autumn and reappears in the spring. Once it was gone there was nowhere to lay eggs. Their dependence on the plant left monarchs in the north with three choices: over-winter as pupae, over-winter as adults, or migrate south away from the bad weather. They had already established a pattern of migration, so they stuck to it. Monarchs do seem to have an inbuilt preference for migration. Populations introduced into Australia by man also migrate to the warmer coastal areas to over-winter. And scientists have found that although keeping them in a warm greenhouse with a supply of milkweed can induce them to maintain a normal sexual cycle, they may still attempt to migrate.

Unfortunately, as the temperate climate spread north, bringing the monarchs and their food plants with it, other changes were happening behind them. The southern United States becomes exceptionally hot during the summer. Temperatures in excess of 40°c are common. In addition to the heat, in many areas it can become extremely dry as well. Milkweed cannot tolerate these conditions, and so dies back. Even if it didn't the monarchs would not be able to deal with the combined heat and lack of humidity. So monarchs in the northern United States and Canada would find themselves effectively stranded, caught between the advancing northern winter and the milkweed-barren south. They couldn't just work their way gradually south as had been done before: even where there were enough flowering plants to feed the adults, there was no food for the young. The adults had to find a way to survive long enough to reach areas where their caterpillars had a chance of survival.

An adult butterfly's short life is entirely geared to mating and producing offspring. The internal production of eggs and sperm is a huge energy drain on the insect's body. Stop the sexual cycle (or go into diapause, to use the scientific term) and the calories taken in as nectar can instead be used to produce body mass in the form of fat and to power long distance flight. In point of fact, migrating monarchs will take as much advantage as possible of thermal updrafts and favourable winds to glide as far as they can, thereby conserving some of their hard won energy.

How far does an individual have to travel? A few do manage to feed, mate and produce young in the southern states, but in late summer and autumn there simply isn't the milkweed available to support many. Butterflies travelling down the eastern part of the continent would find themselves confronted by the Gulf of Mexico. Some do cross to Cuba and the Caribbean islands, but a long journey across open water, with nowhere to land and the possibility of becoming dangerously cold or wet, is far more dangerous to a butterfly than an equivalent journey overland. Most veer west and head through Texas into Mexico.

Having reached the warmth and safety of Mexico, the monarchs could presumably resume their normal lifecycle. However, a few problems immediately become obvious. These monarchs have become adapted to temperate areas, and so would be less able to tolerate the extreme heat of the tropical zones. This excludes them from much of the area in which their ancestors originated. Food would be less of an issue, as flowering plants are common. Their offspring would need milkweed to feed on though, and this would bring them into competition with local inhabitants. If the caterpillars don't get enough to eat, they will be unable to develop into adult forms that can return north in the spring and lay eggs on the newly emerged, unoccupied milkweed plants there. The only option seems to be to wait it out. They must find a place where they can remain safely dormant until the plants begin to rise again in the southern states. Remarkably, they found such a place, and a new generation finds it again each year.

A "normal" peppered moth (© Martin Fowler/Shutterstock. com) and a melanistic one (© Steve McWilliam/ Shutterstock.com)

So far there have been no records mentioning the migration of the monarchs prior to the mid 19th century--with the possible exception of a description from the account of one of Christopher Columbus' expeditions, but that is uncertain. The native people of North America kept no written records and it has been noticed that the butterfly does not appear on the pottery of Mexico prior to the arrival of Europeans. This lack of early information has led some to suggest that the migrations only started within the last two hundred years, possibly in response to the changes caused by the European settlers spreading across North America. It is true that some Lepidoptera can indeed go through this sort of "micro-evolution" with remarkable speed. Consider the case of the peppered moth. Originally a light coloured species with an occasional aberrant dark individual, living in the increasingly industrialised and polluted London of the 18th and 19th centuries caused a rapid change in the population to a mostly dark colour—the better to blend into the soot covered tree trunks and so avoid predators. This sort of change is known as "industrial melanism". In this case, though, it is unlikely that the monarchs' complex pattern of migration and overwintering evolved so quickly. As we discussed above, it probably built up gradually over almost 2my, through the Pleistocene period. However, this is not to say that the activities of settlers had no influence on the monarchs. From the 1850s to the 1870s almost all accounts of monarchs moving in large co-ordinated groups come from the Midwestern prairie states. Starting in the 1880s, though, there are increasing reports of these swarms further east, as far as the Atlantic coast. There is no mystery about what changed in 19th century North America. European settlers, pushing inexorably westward at remarkable speed and in astounding numbers, irretrievably changed the landscape of the continent. Huge areas of woodland were cleared for lumber or farmland (although some 70% of New York State is now officially "forested", almost all of it is secondary, having grown up since the loggers passed through in the mid 1800s).

Across the central plains, some 433 million acres of the intricate and diverse ecosystems of the tall grass prairie were converted to vast monoculture grain farms or livestock ranches. Among the varieties of plants lost during this transformation, of course, were the 22 species of humble milkweed: a poisonous plant with little aesthetic appeal and of no particular use to the settlers (although the fluffy seed

L: an area of Canadian prairie taken over for arable farming (© Jack Cronkhite/Shutterstock.com). It has a beauty of its own, but can't compete as a wildlife habitat with the natural prairie ecosystem, R. (©2009fotofriends/ Shutterstock.com)

parachutes can be used to stuff pillows). Back east, the timber industry had felled most of the forests of the northeast by about 1860, and the Great Lakes area was cleared in the next 30 years. Into the open land that was left moved, among other things, milkweed, and the monarchs followed it.

This shift in breeding areas is not the only effect human activity had on the monarchs. The prairies originally supported at least 22 species of milkweed. However, most of them did not make the transition to the newly deforested areas of the east. Today almost all the milkweed found in the eastern United States is of a single species, *Asclepias syriaca*. All milkweeds contain toxins that the monarch caterpillars can store in their bodies to deter predators. These toxins vary slightly from species to species, and those in *A. syriaca* are relatively mild. Studies have shown that two species of bird and one mouse prey on the butterflies that overwinter in the forests of Oyamel. Most of these butterflies seem to have been feeding on *A. syriaca*, and so their tissues aren't toxic enough to deter the predators.

For the first time in our survey—but not for the first and certainly not for the last in history—human beings seem to have provided the pressure that caused an animal to adapt. All too often our interference has led, directly or indirectly, to the extinction of a species. The reduction in diet available to the monarchs, as well as the gradual destruction of the Oyamel forests, may yet lead to that unfortunate conclusion. But these tiny, virtually defenceless creatures have proved themselves to be both flexible and tough enough to overcome many challenges, in what is perhaps one of the most remarkable adaptations in the animal kingdom.

				12 Dec (200-190mya) Jurassic Earliest Lepidoptera	13 (190-180mya)	14 (180-170mya)
15 (170-150mya)	16 (160-150mya)	17 (150-140mya) Cretaceous Flowering plants spread	18 (140-130mya)	19 (130-120mya)	20 (120-110mya) Gondwana breaking up	21(110-100mya)
22 (100-90mya)	23 (90-80mya)	24 (80-70mya)	25(70-60mya) Palaeocene	26 (60-50mya)	27 (50-40mya) Eocene	28 (40-30mya) Oligocene Climate starting to cool
29 (30-20mya)	30 (20-10mya) Miocene Modern Lepidoptera	31(10mya-present) Pliocene Pleistocene Recent Danaus move from South the Central America (3mya) Repeated Ice Ages in North America until 10,000ya				

Figure 10: butterfly evolution timeline

CHAPTER 5: STRUTTING THEIR STUFF

It sounds a bit contradictory, but we are actually quite accustomed to birds having remarkable features. The great height and speed of the ostrich, the jewel like colours of the parrots and macaws, even the familiar but still enchanting song of the blackbird: somehow we know all these traits are exceptional, but they no longer surprise us. It's just the way birds are.

Sometimes, however, we come across a bird whose appearance and behaviour are so bizarre that they defy calm acceptance. The Birds of Paradise definitely fall into this category. This group of some 42 species—there were even more, but a number have become extinct in recent centuries—are all found in New Guinea, Australia and some neighbouring islands. They are small to medium sized, living in a variety of habitats but mostly in the rainforest. Like so many songbirds, they have a diet of seeds, fruit or small insects. The females and immature males are generally brownish in colour, occasionally speckled or barred but overall very drab. The mature males, on the other hand, flaunt feathered adornments and displays of dancing and calling beside which even the fabulous peacock's fan or the celebrated nightingale's song seem to pale. The largest of the group is the black sicklebill. As its name suggests, the bird is black, but with iridescent blue rimmed fans beside its breast. When displaying he frames his head with these fans, spreads his long tail into a rectangle and tilts sideways on his perch. The King of Saxony's bird displays on a dangling vine, swinging and bouncing to display the two long quills (twice the length of his body) sprouting from his forehead and decorated with sky blue plates. The blue bird of paradise, though far from the showiest of the family, has one of the most remarkable displays: fanning his graceful blue wings around his head, he hangs upside down from a branch uttering the most unnatural sounding call, more like a car alarm than any ordinary birdsong.

Even the name of the family reflects the sense of unreality Westerners felt when in the 16th century they first saw even just the lifeless skins and plumes of the Birds of Paradise. Surely such exquisite creatures belonged rather to heaven, to Paradise, than to earth. The story grew that in life the birds sailed endlessly through the heavenly realms, never needing to alight on the ground (this view may have been assisted, rather pragmatically, by the fact that the traded skins always had the feet removed). In later centuries, when exploration of the area and classification of its wildlife began, the Europeans pulled themselves together and accepted the terrestrial origins of the Birds of Paradise, but still the birds inspired a sense of wonder, as can be seen in the aristocratic and laudatory names some of the species were given: the King of Saxony's bird, Count Raggi's bird, the superb, the magnificent.

Blue (top), Lesser (middle) and red (bottom) birds of paradise displaying. Not to scale(http://animalsspecies.blogspot.com/2011/01/bird-of-paradise.html)

Many birds of the parrot family, for instance (such as parrots, cockatoos and macaws) have brilliant plumage and sometimes crests in a variety of vibrant colours. However, all are fairly simple variations on a theme—it's generally easy to see their family resemblance. The Birds of Paradise are different: even within their own group they differ wildly in colour, feathered accessories, display style and technique, even in their calls.

Eclectus parrots: these are a male and female of the same species. R: rainbow lorikeet. Although this species is much smaller than the Eclectus, you can see the family resemblance (© Allyson Shepard Bailey)

Why? Why did the Birds of Paradise develop these extraordinarily showy costumes: capes, shields, ribbons, wires—in every colour of the rainbow? It seems almost overkill that these birds should have developed such elaborate dances and displays as well. Think of the energy it must take. It is generally understood that in the animal kingdom, males display to attract females. So what made the early female Birds of Paradise decide that what they really wanted in a man was not only a very fancy suit but the ability to hang upside down from a branch and go "boing"?

What, if anything, is actually so special about Birds of Paradise? Is there something unique about them that caused them to develop these extraordinary traits? Or is it their environment, something about the location, geography or climate of New Guinea that allowed or forced the birds to become what they are? The only way to find the answers is to go back and trace the story of both the Birds of Paradise and their homeland, and see if we can identify the significant events.

Greenfinch, a typical small passerine. Note how it's gripping the branch (© Allyson Shepard Bailey)

We had a quick look at the early evolution of birds when we were discussing the emperor penguins in chapter 2 and don't need to go back over their rise from reptiles or very early development. The Birds of Paradise, of course, belong to a very different group from penguins. They are "passerines", or perching birds—birds with three toes pointing forwards and one back, and muscles programmed to grip branches or other perches automatically—which is why they can happily sleep standing on a thin perch. Passerines are also known as songbirds, since they have a specially developed organ in their throat, called a syrinx, which allows them to produce their complex and often very beautiful songs. Since this order contains about 3/5ths of all living birds, assigning the Birds of Paradise to it isn't actually very helpful. Nor, unfortunately, is the fossil record: the light hollow bones and delicate feathers that made the birds such triumphant conquerors of the air aren't robust enough to fossilise well in most conditions. However, molecular and DNA studies are helping to fill in some of the gaps. Here's how the story appears at the moment: It is now thought that the first passerines evolved in Gondwana around 60–55 mya (about 26 December)-this, you may remember, was the supercontinent in the southern hemisphere, that later broke up and became Antarctica, India, Australia, Africa and South America. Certainly by some 55 mya, early perching birds were recognizably distinct. The ancestral passerine was probably a relatively small, drab bird with a monogamous lifestyle: a male would mate with a single female, possibly staying to help raise the young. A short time later, the group began to move out from their place of origin in what would become Australia-New Guinea. They then began what is termed an "explosive radiation", developing new forms and spreading into many new areas very quickly. Presumably along with this radiation came the wide variations in colouring, feeding habit and breeding behaviours that we see today.

For all the different types of passerine and the superficial differences between them—who would guess at the close relationship between a tiny coal tit and a raven, for example—they are actually very similar in their general shape, internal details and lifestyle. So, as with for example whales or cats, the basic pattern was established, and then fairly minor differences developed to take advantage of local conditions. Unfortunately, trying to sort out the family relationships among the passerines has proven to be very difficult. Frankly, the whole thing is a bit of a mess. There are over 5000 different species, and sometimes it's relatively easy to tell which ones are closely related—but sometimes it's not. The different views of various scientists have on occasion led to birds being lumped together into enormous families too diverse to be very useful. New information leads to various species being moved from one group to another.

Left, an aardvark (© Paul Wishart/Shutterstock.com) and a giant anteater (©Allyson Shepard Bailey) Not to scale

Some groupings have been made on the basis of "morphology", that is, the similarities between shapes or functions of specific parts of the birds' bodies. The problem with that is that very different species facing the same problem can evolve very similar ways of solving it. This is called "convergent evolution", and explains why, for example, aardvarks and anteaters have very similar looking faces, with long tubular snouts and very long sticky tongues. Both have evolved to extract ants and termites from inside their nests, and have developed similar tools to get the job done. But they are not related at all. Of course, the morphological grouping of different species by scientists is carried out in a much more sophisticated way, but there are still problems. Notable behaviour has also been used to try and link different species in some way. Until recently, it was assumed that the Birds of Paradise were closely

Bower bird beside its bower. (©Bruce Beehler/Conservation International

related to bower birds, which also live in Australia and New Guinea, in some of the same habitats. They are for the most part rather drably coloured, and do somewhat resemble the females and young, non-breeding males of some of the smaller Bird of Paradise species. Instead of using bright plumage and dramatic performances, they build elaborate, often highly decorated structures called bowers (they aren't nests) to attract females. The inference was that from some ancestral bird, there was a move towards more elaborate courtship, and then a split into two families, with one group turning its energies to display, the other to building. It is a beguiling and reassuringly logical theory, which unfortunately has turned out to be quite wrong.

The comparison of molecular/DNA similarities and differences has produced a slightly different organisation of bird relationships, based on the idea that the more DNA a group shares, the more closely related they are. The technique is somewhat controversial and deciding exactly when one group separated from another is extremely difficult. On the other hand, many people prefer the objectivity of comparing molecular similarities to the dangers of personal bias in the morphological system.

Having all of these problems in mind, this is how the Birds of Paradise are now classified: they belong to a "superfamily" (an extra classification used to group families that are clearly linked together) called the Corvoidea, which includes most of the passerines that now live in Australasia as well as many types that originally evolved in that area but then spread and evolved all over the world in that great explosive radiation. It looks as if the Corvoidea split off from their ancestors around 28mya. Among the families that belong to the group are the Birds of Paradise, bowerbirds, and the corvids themselves: crows, ravens, magpies, shrikes and jays. The corvids, of course, are found all over the world. In fact, it appears that they moved out from Australia-New Guinea and then re-colonised the area later. Birds of Paradise, on the other hand, stayed put.

```
┌──── All other Corvoidea
│
│         ┌──── Drongos
│         ├──── Fantails
│         │  ┌─ Shrikes
│         ├──┴─ Crows and Jays
├─────────┼──── Birds of Paradise
│         ├──┬─ Apostle Bird
│         │  └─ White-winged Chough
│         ├──── Melampittas
│         ├──┬─ Monarch flycatchers
│         └──┴─ Mud-nest Builders
```

Figure 11. The family tree of the Birds of Paradise. I have only included the branches of the Corvoidea that are most closely related to them

Leaving aside their amazing plumage, the other remarkable thing about the Birds of Paradise is their elaborate singing and dancing displays. A number of species display in a group, each one dancing, bouncing or twirling under the generally rather bored looking gaze of a number of females. The Count Raggi's bird males each display on their own individual branch. But despite the number of males displaying, it usually appears that all the females will mate with just one male, presumably because he has the "best" display. An alternative possibility is that he actually has the "best" branch or perch, which may then mean that the displays are actually aimed at rival males, as a means of keeping them away from the most desirable branches.

In any case, this type of behaviour isn't particularly unusual. Peacocks, of course, and other members of the pheasant family, display their beautiful tail feathers and other adornments such as skin wattles. So do grouse and their relatives. They also use the technique of group displays, which is known as a "lek". Each bird has a small area of territory within the lek. Among grouse at least the centre seems to be the most successful or desirable place, and the bird that holds it is most likely to get a mate. Among peacocks the male with the most eyespots is generally the most successful, possibly

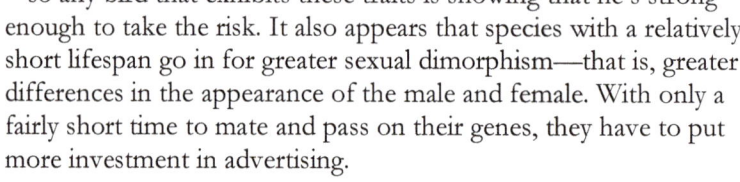

because the older bird (and hence, presumably, the better at surviving), the larger his tail.

There may be a number of reasons why this type of mating behaviour developed. When food is scarce a male will probably have to remain monogamous and help his mate raise their young, so a female will be more interested in a male who can maintain a large territory with lots of food. In any case, the male won't have the time or energy for a fancy display. When food isn't such a problem females tend to choose a mate based on visual signals: they will be raising the young on their own, so all they need is a

Black Grouse lekking (©Steve Ellis/Shutterstock.com)

big healthy male with, presumably, big healthy genes. Being big and bright is dangerous—so is displaying on or near the ground—so any bird that exhibits these traits is showing that he's strong enough to take the risk. It also appears that species with a relatively short lifespan go in for greater sexual dimorphism—that is, greater differences in the appearance of the male and female. With only a fairly short time to mate and pass on their genes, they have to put more investment in advertising.

Ruby throated hummingbird
(©Chas/Shutterstock.com)

There is nothing about the Birds of Paradise that seems to be fundamentally unique: unlike, for example, the hummingbirds which have become extraordinarily specialised in both their bodies and behaviour, a Bird of Paradise without its elaborate plumage could be mistaken for any common thrush or other little songbird. They don't need a very specific diet, and can live in a variety of environments, including the high, cold grasslands of the New Guinea mountains. Even their dancing and singing displays can be matched elsewhere, though perhaps not on so elaborate a scale. So if there is nothing inherently special about the birds themselves, could the answer lie in their environment?

Between 45 and 38mya, the 25th-26th of December, the Australian continent (including New Guinea) finally split away from Antarctica and began to drift north. The earth was growing gradually colder, but it seems that Australia managed to move north just fast enough to keep its climate at about the same temperature for quite a long time, which allowed for the undisturbed evolution of a wide variety of species. Eventually, around 15mya, it collided with the continental plate of Eurasia, and the two land masses pushing against each other caused the mountains of New Guinea to buckle up. It also caused the rise of a number of islands between New Guinea and Asia: close enough together to allow plants to migrate south to the island, but not close enough for mammals to make the journey. This isolation meant that the plants and animals of Australia and New Guinea could follow their own evolutionary path, with only a minimal admixture of characteristics from elsewhere. This is why, for instance, the marsupials such as kangaroos survived and flourished in this area when they mostly died out elsewhere, overwhelmed by the later mammals (The only surviving marsupials outside Australia and surrounding islands are some possums in the Americas).

New Guinea only became separated from Australia about 8000 years ago, as rising sea levels filled the low lying trench between them. So it is an ancient land, but its true isolation is only a very recent occurrence. In many ways it is a unique place, though. It is the second largest island in the world. Lying fairly close to the equator, you would expect it to be very tropical in nature, and indeed it does boast some wonderfully rich tropical rainforest. Although physically closer to Australia, the climate of New Guinea is for the most part more like that of SE Asia. In different parts of the island there are humid swamps, coastal savannahs and high altitude cloud forest. The tropical climate allows for a year-round supply of fruit and insects for larger creatures to feed on. However, the range of mountains running along the spine of the island, and the upper plateaus, are high enough to have a much more temperate climate, and up here it can be cold, even snowy, with sparse vegetation and there isn't the constant abundance of food found at lower levels. The landscape of the whole island is fractured by mountains, rivers and other features into a number of separate areas. These pockets are isolated enough to limit the flow of animals from one to another, and without the continual mixing of genes caused by migration each group develops into separate species (This fragmentation affected the human populations as well: there are over 1000 different languages native to the island). Birds of Paradise have been diversified in the same way. Most species inhabit a fairly small area, usually in a single mountain group and within a fairly narrow altitude range.

 The long standing physical link means that north western Australia and New Guinea share many animals and birds. For example, a few Bird of Paradise species live in Australia (though interestingly they are among the less showy and dramatic of the group). There are 189 native mammals on New Guinea, two "monotremes" (egg laying mammals, the most ancient branch of the family) and the rest marsupials, except for some bats and rats that managed to fly or island hop from further north. There were larger predators such as the Tasmanian wolf a few million years ago, but they are extinct now, having disappeared from New Guinea around 3000 years ago. Like New Zealand, another large island community without large mammalian predators, New Guinea has become a paradise for birds. The island is home to 578 species, of which 324 are native.

Birds of Paradise, as we have seen, are quite closely related to the Corvids: crows, jays, ravens and so on. Corvids are for the most part unspectacular birds, often plain black, with harsh, unmusical voices. Their most notable characteristics are intelligence and adaptability. They can survive in many unpromising habitats, often living on carrion or, more recently, on human rubbish. Jays hide nuts and return to find them months later. Carrion crows in Japan and California drop nuts at busy intersections so that the cars will crack them as they drive over, then hop out to eat the nuts while the lights are red. Caledonian crows not only use but make tools, a skill once arrogantly believed to belong only to humans. Corvids have even made their way into human mythology: Raven the "trickster" appears in several cultures.

This sort of intelligence can be of great use when it comes to evolution, helping a species to overcome difficulties and survive where rivals with less developed problem solving skills would perish. This is the basis of natural selection, "survival of the fittest". But New Guinea really doesn't offer much in the way of challenges. The climate is benign, the land provides a virtually constant supply of food, there are no large mammals such as monkeys to compete for resources and virtually no predators. In fact, most of the pressures that cause the process of natural selection to operate are almost non-existent. In harsher environments, merely surviving to sexual maturity is a sign of success and any animal that does so can be assumed to have some genetic quality that helped it and that can be passed onto its offspring. That individual is a good catch, sexually speaking, and probably has a good chance of mating.

A successful Bird of Paradise doesn't have to be particularly strong or clever. By and large they don't have too many problems surviving until they are mature enough to breed. If all of them did so, of course, the population would grow beyond sustainable levels, with too many birds for the available food supply, and whole populations could starve. So the other side of the evolutionary coin, sexual selection, has become much more important. The idea of sexual selection is that an individual gets chosen to mate and produce offspring if it exhibits traits that are considered useful or desirable by the opposite sex, or sometimes if it can fight off the competition of its own sex. Stags and bull elephant

L to R: Astrapia ribbon tailed and King of Saxony and Raggiana Birds of Paradise.
(http://animalsspecies.blogspot.com/2011/01/bird-of-paradise.html) Not to scale

seals fight one another, with the victor claiming the right to mate with a harem of females. The leks and other displays used by pheasants and grouse, that we discussed earlier, are all part of this same process.

Without much need to overcome any physical hazards, the Birds of Paradise have evolved to rely almost entirely on sexual selection to ensure that only the fittest birds breed. With no problem finding food, there is no particular need for territory—interestingly, the Macregor's bird of paradise, which lives at the highest altitude of any Bird of Paradise species, in colder areas where food is less abundant, is relatively drab and monogamous and doesn't have an elaborate ritual display. It also becomes easier to spend energy on growing colourful, extravagant and, perhaps most importantly, virtually useless feathers. With no predators to worry about, displaying on or near the ground becomes safe and so the birds can concentrate on long and extravagant songs and dances. There are no particular pressures to move, either to find food or to escape danger, so each population tends to stay in its own little area. This means there is less homogenisation of the different groups, although in fact they can hybridise, or cross breed, quite easily.

In a manner of speaking, the astounding beauty and extravagance of the Birds of Paradise is actually the result of boredom. They simply have no other challenges to overcome. With no need for the corvid cleverness, they have turned to using their inherited gift of adaptability for display purposes, diversifying their appearance and performances to a remarkable degree. Those early females obviously decided they wanted a man who could hang upside down and go "boing" because he seemed a slightly better bet than one who just went "ta-da!"

22 Dec (100-90mya)	23 (90-80mya)	24 (80-70mya)	25(70-60mya) Palaeocene	26 (60-50mya) 1st Passerines in Australia/New Guinea	27 (50-40mya) Eocene Passerine explosive radiation Australia/New Guinea split from Antarctica, move North	28 (40-30mya) Oligocene

29 (30-20mya) Corvoidea split	30 (20-10mya) Miocene Australia/New Guinea collide with Eurasia	31(10mya-present) Pliocene Pleistocene Recent New Guinea separates from Australia Tasmanian wolf and other predators extinct in New Guinea

Fig 12: Bird of Paradise evolution timeline

CHAPTER 6: HANGING OUT

The sloth has the dubious distinction of being the only animal to be named after a human sin (though the wolverine has sometimes been known as "the glutton"). In some ways this seems unfair. After all, there are animals that hibernate for six months or more each year, and a well fed big cat can sleep 20 hours a day. But when a cheetah awakens it can run faster than any other animal, and a hunting polar bear, having come out of hibernation, can travel many miles in search of its prey. The sloth, on the other hand, regularly sleeps and rests for 20 hours a day, usually travels an average of no more than 24 metres a day and can achieve a maximum speed of 37 metres per minute through the trees—and only 4.5 metres per minute on the ground. Their eyesight is relatively poor, and they focus on objects with difficulty, especially close up. Their hearing is also poor, or at least selective: while they will respond to a sound they consider important, such as their young crying, they will hardly react at all to a gunshot, presumably because they don't hear well enough to find it startling, and they simply aren't interested in it. Their sense of equilibrium is rudimentary, as is their sense of touch: a falling sloth will make little attempt to right itself; and if you tap one it will apparently not even be sure which part of its body you touched. They have an extremely low proportion of muscle to bone and very little subcutaneous fat (that is, fat deposited under the skin). They appear, in fact, to be extremely inefficient creatures. Yet, sloths are one of the most abundant large mammals in the Central and South American rainforests. How could an animal seemingly lacking in most of the skills, senses and physical qualities necessary for survival have evolved and continue to flourish for some 18 million years? Or to look at it in a more positive light: what special qualities do the sloths have that allow them to survive and indeed flourish, despite what appear to be great disadvantages?

Let's first examine the sloth in more detail. There are actually two different varieties: the three-toed, or *Bradypus*, and the two-toed or *Choelepus*. The three-toed is generally slightly larger, but both are roughly the size of domestic cats. Both types have rather pear shaped bodies with a large abdomen to accommodate the digestive apparatus needed to digest their diet of leaves. Their bodies are covered in long coarse fur with a thick soft undercoat. Famously, their fur grows from their belly and hangs down their back, the reverse of the pattern seen in other mammals, and in some species a type of algae grows in grooves on the hairs, giving the animal a greenish appearance. Moths live in the sloths' fur and may in fact feed on the algae, laying their eggs in the sloths' dung for their larvae to feed on.

Sloths have small round heads, short noses, small ears hidden in their fur and small eyes. Their limbs are relatively long and slender. An additional minor oddity of the sloths is that their toes are bound together with skin, rather like the webbed feet of a duck, so their long curved claws are sometimes mistaken for digits. As we mentioned before, they have poor eyesight. The retina at the back of our eyes is lined with two structures called rods and cones, allowing us to see both shapes in low light and colours in daylight. Sloths have a few cones but the majority of their light sensing cells are rods, so their eyesight is better at night—though their pupils can contract very tightly, which is very useful in bright tropical sunlight. The sloths' skeleton has two adaptations that seem to act to their advantage: the three-toed sloth has 9 vertebrae in its neck, which allows it to turn its head further than most animals (the two-toed has 6 or 7, which is more standard among mammals); and both varieties have extra surfaces

Three toed sloth (*Bradypus*, © worldswildlifewonders/Shutterstock.com) and R: two toed (*Choelepus*. ©Grigory Kubatyan/Shutterstock.com)

on their vertebrae for muscle attachments. Having this extra strength and flexibility allows them to stretch from a horizontal to a vertical support while holding on with just their hind limbs. An animal that wants to move through the trees but has neither the strength nor the speed to jump or swing would obviously benefit from the added flexibility and control given by these adaptations.

Sloths have a fairly small number of teeth, which lack the hard enamel that protects most animals' teeth from wear. The teeth are simple cones when they first emerge from the gums but develop distinctive "cusps" or points as they become worn. They don't have the chisel shaped incisors that allow rodents to gnaw their way through so many materials—our front two pairs of teeth are incisors—nor the long pointed canines of a predator (we have canines as well, though they aren't as impressive as those of true carnivores).

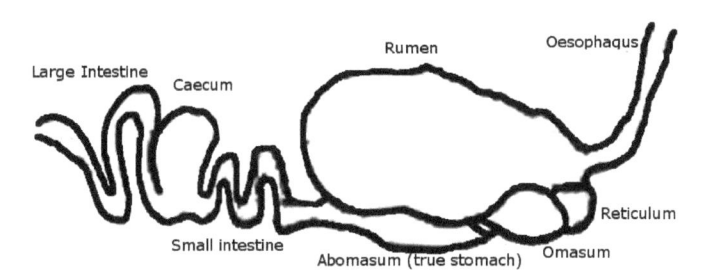

Fig 13: The ruminant digestive system. Food is broken down in the rumen by bacteria. The reticulum is used to bring "cud" up to be re-chewed.

An animal that lives almost exclusively on leaves doesn't need very specialised teeth, but it does need a very efficient digestive system if it is going to extract enough nutrition from its diet to stay healthy. Leaves are notoriously low in calories and nutrition—which is why people who are trying to lose weight eat so much salad—but high in cellulose fibre, which is difficult to break down and digest. The most familiar leaf eaters (or "folivores") are the grazers, such as antelope horses and cattle—technically, grass blades are leaves, and of course some grazers such as goats and deer also browse on the leaves of other plants. Many of them solve the problem of extracting enough nutrition from their meals by effectively eating their food twice: lumps of partially digested grass are brought up to the mouth for a second chewing and then pass into the stomach for final digestion (This is what cows do when they are "chewing the cud"). Rabbits excrete the remains of their food, then eat the droppings again so that the material can be digested further and more of the nutrients can be absorbed. Cows and other ruminants also sustain colonies of bacteria in their stomachs. The bacteria happily munch their way through the otherwise indigestible cellulose and allow their hosts to absorb more of the nutrition in the leaves they eat.

Small folivores such as rodents tend to excrete the larger, harder to digest particles while retaining the smaller ones from which they can get some benefit. Sloths, contrary to popular belief, don't spend most of their time hanging suspended from a branch. A recent study found that three-toed sloths spend some 50% of their days sitting up. In this position the larger particles of food in the sloth's stomach float to the top and, because of the design of their digestive system, will pass through the gut and be excreted more quickly than the smaller but heavier pieces that sink to the bottom and so remain in the system being digested for longer.

Sloths eat leaves from a variety of trees and plants, though by no means all that are available to them. The two-toed has a much wider palate than its three-toed cousin (which is why they survive well in captivity, while the three-toed type doesn't). Neither species necessarily eats those leaves that are most common in their home range, so clearly they are making definite choices. They seem to learn their food preferences from their mothers, and mainly stick to the same types of leaves they were fed as babies. Leaves often contain toxins as a defence, so any animal that relies on them must develop some means of ensuring that their meals are safe. Sloths seem to prefer young leaves, which have fewer toxins and a higher protein content than older, tougher ones. They are often partial to Cecropia leaves, which have fewer toxins than some other types.

David Attenborough once jokingly remarked that little was known about sloth's behaviour because no researcher could put up with the tremendous boredom of studying them. However, some patient investigators have managed to piece together some of their life cycle. Sloths for the most part live solitary lives, generally in lowland tropical rainforest although the two-toed can also be found in higher altitude cloud forest. Although they can swim fairly well, on the ground their limbs are so weak they can only move at a slow flailing crawl. (Baby sloths, oddly enough, are slightly better at supporting their own weight and moving around). In the trees they move by using their long strong claws as hooks and

Top: a sloth struggling to move on the ground (© Worldswildlifewonders/Shutterstock.com) and bottom, hanging from a branch (© Jacinto Yoder/Shutterstock.com)

pulling themselves from branch to branch. When hanging suspended the claws can lock around the branch extremely tightly and with no effort from the sloth's muscles. In fact, a sloth in this position, if shot and killed, will continue to hang onto the branch, and the hunter must climb up to release it.

Their time is divided among several activities: feeding, grooming themselves, resting and moving. Resting takes up most of their day (one study found that the three-toed sloth rested some 73% of the time). As mentioned above, they actually spend more time sitting up than in the famous hanging posture, but in either case they can be either huddled up or sprawled out, depending on whether they can expose themselves to the sun. Although it would appear that hanging is easier than sitting up, the sloth's heart rate is actually higher when it is hanging. It has been suggested that this posture exposes their belly to the sun and helps to speed the production of enzymes used in digestion.

Sloths will move within or between trees in order to find food, changing trees every one and a half days or so: the longer an individual tree is fed on, the more toxins it will produce as a defence. So it pays to move to a new site, where the tree isn't yet geared up to fight off attackers. They also come down to the ground to urinate and defecate, perhaps once or twice a week. An animal that can only crawl not much over 4 meters a minute seems to be just asking for trouble returning to the ground in this way, and indeed many sloths are killed, often by jaguars , while they are on the ground. Few tree dwelling creatures bother with such niceties, so why do sloths? Perhaps the dung deposits act as territorial markers, or perhaps the behaviour is a remnant from a time when the sloth's ancestors were ground living, and a bit quicker on their feet. The latrines are used by a number of sloths and are extremely smelly, even to insensitive human noses, so this could be where one sloth has a chance of meeting another in order to mate.

Because their bodies don't regulate temperature well, the sloths' thick fur is extremely important, keeping the animal from becoming dangerously chilled during heavy rain or colder weather. They seem to spend between 5 and 6 percent of their time grooming, though mostly on their front: presumably they lack the agility to deal with the harder to reach areas.

Although they are mostly solitary, sloths do come together to mate, though little is known about how they court or choose a mate. They become sexually mature around the age of three. The female raises her single youngster alone. The baby stays in constant physical contact with its mother for about six months, long after it is weaned. It presumably uses this time to learn which foods it can eat. Eventually the mother, rather than driving her offspring out of her territory so it can set up one of its own, moves into a different part of her range leaving the young one in command of the familiar area in which it was raised.

Interestingly, baby sloths, as well as moving more quickly and being better at walking than their parents, also have better eyesight. They lose these skills as they mature.

How does an animal with no speed or weaponry, poor senses and a slow reaction time defend itself from predators? Although the sloth can move relatively quickly when threatened (though nowhere near as fast as a cat, for example, which is of a similar size), and has long formidable looking claws, which it will wave while emitting a mildly threatening hiss, it is not much use at either fighting or fleeing. It relies instead on what is known as "cryptic" behaviour, meaning simply "hidden" (the original meaning of the Greek word from which ours is derived. Only later did it take on the additional concept of something mysterious or confusing). Motionless or moving very slowly, huddled against a tree trunk or hanging from a branch, the sloth with its inconspicuous fur can be virtually invisible. The algae that colonise its fur can add a greenish tint, adding to the camouflage and perhaps partially masking its smell as well. Many predators rely on identifying movement and outline to spot their prey, so the sloth's strategy works quite well.

So why are sloths, in fact, so slothful? Clearly their environment must have something to do with it. The warm, rich tropical jungles of central and South America offer abundant food and little in the way of hazardous weather. Yet the qualities that make these areas so attractive ensure that competition is fierce and unrelenting, and so rainforest dwellers are generally no more indolent than animals living in less benign ecologies.

Some arboreal folivores: L; howler money (© Naaman abreu/Shutterstock.com), centre: tree hyrax (© Simon_g/Shutterstock.com), and R: koala (© Allyson Shepard Bailey)

Is there something physically inherent in the sloths that makes them so slow? Some scientists have looked quite carefully into the details of their physiology, digestion and other bodily processes to try and answer that question. Sloths have a much lower metabolism than other mammals of a similar size, even allowing for the fact that they have relatively little muscle mass to provide power for. Their body temperature is low and they are not very good at regulating it internally, which is why they tend to move into and out of the sun, and higher and lower in the trees, if they need to be warmed or cooled. Despite their leaf diet they have a fairly high level of cholesterol in their blood, which indicates that their thyroid activity is rather low. Their lungs have a relatively small surface area for collecting oxygen, so that they extract only two percent of the oxygen from the air they breathe (we should not feel smug: we are not much more efficient, obtaining an average of just four percent). On the other hand they are able to stop breathing for up to 40 minutes if necessary. Their heart is fairly small and beats slowly, but their blood pressure is rather high, that being the only available mechanism to keep the blood moving around their system. Although sloths have a very powerful grip, their muscles contract only slowly, and the speed of impulses along their nerves is rather slow as well.

Are they slow because they have relatively little muscle and a feeble heart rate? Or did these adaptations develop because a lethargic animal doesn't need a strong heartbeat or powerful muscles? An underactive thyroid can lead to sluggishness, but the sloth's is not exceptionally low.

We mentioned earlier that sloths escape predators by being hard to spot. Could this be linked to their very slow movements overall? It seems unlikely: many prey species, such as mice, instinctively freeze to avoid the attention of a predator, but can then skitter off with great rapidity.

It is generally assumed that the sloths' low nutrition leaf diet is responsible for their sluggishness. Certainly some other folivores are less active than related species with a higher energy diet: howler monkeys, for example, spend much more of their time resting in the trees than other monkeys, while their oversized guts digest the leaves that form the largest part of their diet. But they still retain the social activities of other primates, calling loudly to establish territories and maintain links within their social groups. Grazing animals may have to spend most of the day eating, but if necessary some species can outrun all but the fastest of predators. Clearly eating leaves doesn't necessarily mean giving up all speed of movement or reaction, but it does impose some definite restrictions, particularly on an animal that wants to spend its life in the trees. You must be small enough to move around without the branches breaking, but large enough to maintain a digestive system that can detoxify, break down and extract the energy from the leaves. There are folivores in all different orders of mammals including marsupials, primates and even the tree hyrax, an unlikely relative of the elephant that we mentioned in Chapter 1. All tend to move more slowly than their relatives with more energy rich diets, and are generally rather smaller. They have dense fur to help limit the energy they spend on heat regulation and tend to have small litters of young: such sluggish parents couldn't cope with large numbers of active babies. Folivores in general live very close to the limit of their "energy budget", just managing to collect enough energy from their food by reducing the demands of high energy systems.

Sloths, however, have taken these adaptations to the extreme. Presumably they became folivores because that was an unoccupied niche in the local environment. But that still doesn't explain why they should have become so much slower than is apparently necessary, as well as generally being so underperforming. Obviously some ecological pressure drove them along this road.

The two varieties of sloth, two-toed and three-toed, are in fact not very closely related. Each is the modern representative of a different ancient family, but for some reason both elected to follow the same evolutionary route. Perhaps tracing the evolution of these creatures will help us to understand them better.

Sloths belong to an order of mammals known as Xenarthra, which also includes the anteaters and armadillos. At first glance there doesn't seem much to link these three types of creature, other than the fact that they don't seem to resemble anything else. However, they do share a number of features. Some are internal, such as the extra articulations in their lumbar vertebrae and extra surface area on the scapula or shoulder blade: these give the animals extra stability and added power in the forelegs for

Top L: giant anteater (© Allyson Shepard Bailey) R: tamandua anteater (© Karel Gallas/Shutterstock.com) Bottom L: three banded armadillo(©Bonnie Fisk/Shutterstock.com) R: nine banded armadillo (© Ron Kacmarcik/Shutterstock.com). Note the plate on the three banded armadillo's face: when it curls up this plate fits beside one on its tail, forming a solid ball.

digging or climbing. Their brains are relatively small and their teeth are reduced; the anteaters, of course, have none at all. Perhaps the most visible connection among these animals is their thick, strong curved claws. Anteaters' claws are so long, in fact, that they walk on their knuckles with the claws turned in and up.

Otherwise, the three families are quite different. Sloths, as we know, are small, furry and live almost entirely in the trees, mostly eating leaves. Anteaters range in size from the diminutive silky anteater, 16-21cm long and weighing no more than 275 grams, which is specialised for life in the trees, to the famous giant anteater, which can be up to 2m long and weigh 20-40kilos. These live on the plains, breaking into termite mounds with their long powerful claws. Although they tend to walk at an amble, they can run fairly quickly, and their hug is exceptionally strong: there are stories of anteaters killing jaguars in this way.

There are about a dozen species of armadillo, ranging in size from the three-banded, which can roll itself into a ball about the size of a grapefruit, to the giant, which is roughly the size of a large bulldog. All are protected by bony plates covered with horn over the upper parts of their bodies and heads. The armour isn't a single solid piece like a turtle's shell, but divided into plates connected by flexible skin. This allows them to bend their bodies, some rolling up like a hedgehog to protect themselves. All are powerful diggers, happily digging burrows to sleep in or to find food. Armadillos have a more varied diet than their Xenarthran cousins, eating fruit, carrion and insects. The giant lives almost entirely on insects, tearing apart ant and termite nests with its large claws and sweeping them up with an exceptionally long sticky tongue. This adaptation clearly shows the link with anteaters.

The mammal line that would lead to the Xenarthrans seems to have developed around 75-80mya. The earliest true Xenarthran appeared in South America some 60mya, near the end of the 24th of December on our calendar, so they are relatively modern animals. The first members of the family were smaller than a common armadillo today, but looked quite a bit like them. At this time the world was much warmer overall than it is today, with much of the land covered with tropical vegetation. The gradual movement of continents, breaking up the great mass of Gondwanaland, had left South America isolated, so for some 60 million years the plants and animals there, including the Xenarthrans, were able to evolve in their own particular way, unaffected by developments elsewhere.

Utaetus, the earliest known Xenarthran (© Eleanor Loughlin)

By the early Eocene, some 40 mya or 24 December, the earliest armadillo-like Xenarthran family had developed another branch, which would lead to the sloths and anteaters. Although the fossil record for this period is so far rather poor, it seems clear that the group abandoned the protection of armour for the increased mobility and temperature control offered by the mammal's distinctive coat of fur.

Like many other mammals, some of the Xenarthrans took advantage of the gaps in the environment left by the disappearance of the dinosaurs to grow to enormous size. Glyptodont, a now extinct cousin of the armadillo, reached the size of a small car, trundling across the grasslands under its huge domed shell. It was the most heavily armoured animal to ever exist. One of these shells has been excavated showing indications that humans may have used it as a living space or shelter. Some of the ancestral sloths followed the same route: some members of the family Megatheridae (a name that comes from two Greek roots simply meaning "big animal") were the size of elephants. The teeth of at least some members of this family seem to indicate that they were browsers; and fossilised dung has shown that they ate at least seven varieties of plant. These were, obviously, ground living animals, and indeed are commonly known as ground sloths, to distinguish them from their modern tree dwelling descendants. They moved on all fours, resting on the backs of their front feet with their claws turned in

Glyptodont, the giant ancestor of today's armadillos(© Eleanor Loughlin)

and up, like a modern anteater. Also like the anteater, they sported a long bushy tail that the modern sloths have lost. The tail may have helped them to balance when they sat back on their hind legs to reach up to leaves on the branches above them. An animal the size of an elephant, of course, would find it very hard to live on insects or even carrion, and so these giants seem to have been purely vegetarian, perhaps already mostly leaf eaters. They seem to have got themselves locked into an ascending spiral, growing larger to reach higher leaves, then having to reach higher and higher to collect more leaves to feed their bigger bodies.

A megalonychid, one of the giant ground sloths. (© Eleanor Loughlin)

However, the fossil sloths diversified into a number of different types occupying different ecological niches and eating different things. Most seem to have been vegetarian, though some may have been omnivorous scavengers like Miocene armadillos. They were terrestrial, semi-arboreal and fully arboreal. All showed a large range of movement in the elbow joint, though none was as extreme as in modern sloths.

By 12 mya in the Pleistocene, 29 December, there were over 40 different genera of sloths spread all across the Americas from Alaska to the southern tip of Argentina. They varied in size from the elephantine Megatheridae to smaller tree dwelling species. Some of the giants may have survived until 10,000 years ago, perhaps hunted by the first human settlers of the Americas as they spread southwards. However, as the Ice Age brought its dramatic climatic shift and habitats and vegetation changed, many of these early sloths disappeared. The giants presumably fell prey to the same conditions that destroyed the giant herbivores such as the mastodon, and the sabre-toothed predators that pursued them. But some of the smaller, tree dwelling species survived. Perhaps their life in the trees protected them from new predators that moved into their home area, or a diet of leaves proved more consistently available than whatever their ground living cousins were eating. It has been suggested that climbing trees helped them to escape floods, but since the advancing Ice Age led to a drier climate, this seems rather less likely. One thing is clear: the two types of arboreal sloth that survive today, the two-toed and three-toed, are only rather distantly related and evolved from different ancestors. So the "decision" to adopt their apparently rather improbable lifestyle was taken independently at least twice, and therefore must have had some clear advantages.

Arboreal yes, but why develop such odd features? As we have seen, other folivorous tree-dwellers also tend to be rather more sluggish than other members of their families, so clearly that is a feature of this lifestyle. The ancestral ground sloths were, of course, very large, and unlikely to be bothered by many predators, and they were presumably quite slow moving. In fact, no modern Xenarthran is particularly quick. So unlike, for example, howler monkeys, modern arboreal sloths are not descended from very active ancestors, and would have no reason to want to acquire speedy habits. This is probably the most important starting point in the development of the modern sloths' habits. You don't need to be quick to hunt down leaves. No Xenarthran even today is fast enough to evade predators, and they rely instead on hiding or the protection of armour. So the slower a sloth moves, the more inconspicuous and therefore the safer it is. This is another important difference between sloths and other folivores: most of them must spend the majority of their time eating, in order to take in enough nutrients to fuel their high energy life style. Once the sloths reduced their activity level, then they wouldn't have to spend as much time eating, even on a leaf diet. But then, if they reduced their intake of food, particularly when the food they relied on was quite low in nutrition, they would have to greatly reduce the demands they made on their bodies, because they would have so little fuel available to meet those demands. In any case, if you're going to live on leaves, you have to spend a lot of time just waiting for them to digest. Having elected to follow a very slow moving, "cryptic" lifestyle, the sloths had no need for big energy-hungry muscles, so they allowed them to dwindle away as far as it was possible and still allow some purposeful movement. They don't hunt their food, or run away from predators, so they don't need high energy senses to detect either food or danger, nor do they need to feed a quick responsive brain to direct them. By moving up or down in the canopy, or into and out of sunlight, to warm or cool themselves as needed, they also reduced the need to use precious fuel to regulate their body temperature. Thermoregulation, as this process is called, is extremely expensive. That is why a "cold blooded" reptile, using the external environment to warm or cool itself, can get by eating much, much less than a similarly sized mammal.

It seems that we have been misjudging sloths. Far from being extremely inefficient creatures, they are in fact very well designed, having pared down their physiology and lifestyle to the absolute minimum. There is nothing superfluous or wasted in a sloth. They have reduced their energy needs, intake and expenditure just about as far as is possible. In these days of climate change, and dwindling resources, when we are all concerned to reduce our consumption, recycle as much as possible and shrink our "carbon footprint", we should probably be saluting the sloth as an excellent example to follow!

15 Dec (170-150mya)	16 (160-150mya)	17 (150-140mya) Cretaceous	18 (140-130mya)	19 (130-120mya)	20 (120-110mya) Gondwana breaking up, Australia and Antarctica still connected	21(110-100mya)
22 (100-90mya)	23 (90-80mya)	24 (80-70mya) Xenarthran ancestor	25(70-60mya) Palaeocene	26 (60-50mya) 1st true Xenarthrans	27 (50-40mya) Eocene	28 (40-30mya) Oligocene Split between armadillos and sloth/anteater branch
29 (30-20mya)	30 (20-10mya) 20 genera of sloths all over the Americas, including giants	31(10mya-present) Pliocene Pleistocene Recent Giant sloths become extinct, small arboreal species survive				

Figure 14: sloth evolution timeline

CHAPTER 7: BITING OFF MORE THAN THEY CAN CHEW?

Milk snake (©Matt Jepson/Shutterstock.com)

Cape Cobra (© Ecoprint/ Shutterstock.com). The cobra is truly venomous. The harmless milk snake mimics the colours of the venomous coral snake to make itself appear dangerous.

Certain types of animals seem to inspire more intense and more polarized emotions than others. Snakes are one such group: some people are utterly fascinated by them, while others find them completely terrifying. While I am certainly wary of the dangers some snakes can pose, I personally lean more towards the fascinated. Snakes are so different from most of the animals we encounter in our normal lives. Perhaps the most obvious difference is the fact that they are completely without limbs. But unlike worms, they have an internal skeleton anchoring strong muscles: handle a snake and you are immediately aware of the pure power turning in your hands. Unlike eels, they have a hard, dry skin, often beautifully coloured and patterned. Their unblinking stare—that is a literally true description, snakes have no eyelids—is reputed to hypnotise their prey: whether this is true or not, many people find it unnerving. They have no true ears, often perceiving movement by vibrations transmitted to their jaws. They "smell" using their flickering forked tongues. Their eyesight is unremarkable, but some species supplement it with heat sensors. Most of them kill their prey in one of two ways. The first is constriction: that is, wrapping their body around the prey and squeezing until it suffocates. Other snakes are venomous, injecting venom into their victims through their hollow, hypodermic needle-like fangs. Few vertebrates carry venom and even fewer actively hunt with it, so this is another characteristic that sets snakes apart.

But perhaps the most unique feature of snakes is the way they eat. Although a snake's skull and jaw are quite different from our own, the basic pattern is the same: a relatively solid upper skull, with some gaps or "fenestration" between the bones and teeth lining the bottom edge, and then a lower jaw hinged to the skull on either side, with more teeth arranged so that they can work with those in the upper jaw. (A snake's skull is in fact much more flexible than our own, and also has more fenestration—we'll discuss this in more detail a bit later). Many tetrapods (animals with 4 limbs—really most vertebrates are or were tetrapods) have a bone between the base of the lower jaw and the skull

A Western diamondback rattlesnake with its jaws at full gape (© Audrey Snider-Bell/Shutterstock.com)

called the quadrate. It forms part of the actual jaw joint (In mammals, the quadrate bone has migrated and now forms part of the middle ear). In the case of snakes, the quadrates have become enlarged and now slant slightly backwards. In effect they have become loose struts attaching the jaw to the skull. In some snakes the cartilage joining the hinges of the jaw is particularly flexible, allowing them to stretch their jaw vertically much further than we can. The jaw itself also has other special features: at the point of the chin, where we have solid, continuous bone, snakes have another "joint", a flexible connection that allows them to spread the two sides of their jaw apart (Humans have a similar connection, or "symphysis", at the front of the pelvis, which loosens up, sometimes painfully, during pregnancy to give slightly more room for the baby to be born). At either side of a snake's jaw is another "joint", which allows them to bow out the sides of their mouth and engulf even larger prey.

A grass snake eating a fish—notice how the prey is actually larger than the snake's own head (© Dinkey Creek/ Shutterstock.com)

Very few snakes dismember their food or bite pieces out of it. Whether they kill using venom or are constrictors, squeezing their prey to death, nearly all snakes then eat their victim whole, engulfing it gradually with their head and body and then lying torpid until it has been digested. Some snakes can bring down prey much larger than themselves, and swallow food whole that is larger than their own heads. Taking prey larger than yourself is fairly common in the animal kingdom: many big cats, some dogs, even weasels and stoats do it. But as a general rule they then eat the victim progressively, biting or tearing off mouthfuls. Some predators may swallow their prey whole, but in that case the victim will almost certainly be much smaller than the predator—such as an owl eating a mouse or vole. As anyone knows who has eaten a large meal quickly, it's usually best to spend some time resting while you digest. A large snake that has just swallowed a meal equating to 25% or more of its own body weight (some can even take in food amounting to 150% of their weight!) will have to spend up to a week or more staying more or less motionless while its system deals with the meal. During that time it will be vulnerable to any other predator.

Clearly there must be some advantage to this behaviour, but at the moment it's hard to see what it could be. So why did snakes evolve their peculiar jaw structure and feeding behaviour? In order to understand this we need to go right back to the beginning of their evolution and trace when and how the physical characteristics developed.

Snakes, of course, are reptiles, and reptiles, loosely speaking, evolved from amphibians. People often confuse lizards, which are reptiles, with salamanders, which are amphibians, and it's easy to see why. Both have very much the same body plan: a long slender body without too much difference in size at the neck and tapering evenly through the tail (unlike most mammals, where the tail is noticeably more slender than the animal's hindquarters). Their four legs splay out on either side of the body, though the lizards hold their bodies a bit higher than salamanders can. They move in a distinctive pattern, moving the fore leg on one side and the hind leg on the other forward together, bending their whole body back and forth in an S shape as they wriggle along.

Above, a lizard
©Renate/Shutterstock.com)
Below, a salamander
(©Gorilla/Shutterstock.com)

However, there are some very important differences between the two groups as well. Salamanders have smooth, cool, moist skins whereas lizards have hard, dry skin, sometimes thickened into rough, heavy armour or scales. You can also see a big difference when they breed. Amphibian eggs are soft and shell-less. They must be laid in water, or at least in very moist areas, because if they dry out the developing embryo will die. Reptile eggs, on the other hand, have a waterproof shell (it's leathery, rather than hard and brittle like a bird's egg). This means the baby reptile is effectively enclosed in its own personal pond. So the eggs can be laid almost anywhere, even in the desert, without danger of drying out.

Waterproof skins and waterproof eggs are the two main innovations that characterise the evolution of amphibians into reptiles. These new features allowed the reptiles to expand into habitats in which the amphibians would never survive, and therefore where there would be relatively few competitors or predators. And just as we would expect, the reptiles changed and diversified as they spread into these new habitats and lifestyles. We don't need to go into the earliest part of the story in detail, but these are the basic facts: the earliest known hard shelled egg dates to roughly 300mya (around 1 December), and the earliest fossil that can definitely be considered a reptile came from deposits that are even slightly older. This was a creature named "*Hyonomus*", a small (about 20cm long) lizard like creature with a heavy solid skull. The reptiles evolved first on Pangaea, the early "supercontinent": at this point all the continents we know today were joined together in a single landmass. It was centralised on the equator so that the overall climate was generally warm, humid and lush: a perfect reptile environment. During

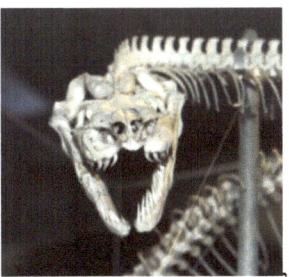

From L to R: a snapping turtle skull (© Robynrg/Shutterstock.com), solid bone from the eye sockets to the back of the skull; a horse skull (©Henk Vrieselaar/Shutterstock.com), with a single opening behind the eye socket; and a snake skull (@ Aaron Kohr) with 2 openings on each side.

the first hundred million years after the emergence of reptiles, they divided into three main lines, distinguished by the details of their skulls. The Anapsids had no openings or fenestration at the sides of the skull. Once there were many Anapsid reptiles but they are now all extinct. Today the only reptiles with this type of skull are turtles and tortoises, but they may have evolved this feature independently. The group eventually gave rise to two more: the Synapsids, the branch that would eventually lead to the mammals, evolved a single opening called the temporal. You can find your own temporal openings on either side of the head just behind the eyes. It's where the jaw muscles pass through to the top of the skull. The third group, the Diapsids, gave rise to the dinosaurs and also modern lizards and snakes. Instead of one opening on each side of the skull, the Diapsids have two. Among other things, having two openings allowed the Diapsids to have stronger jaw muscles and open their jaws more widely than the other groups. We can ignore the mammal and turtle lines for the moment. Over the next 130my the Diapsid branch divided further into crocodilians, pterosaurs (flying reptiles, which sadly are now extinct—birds actually evolved from the dinosaur line), a group known as rhynocephalia or "horned heads", which only has two living members, the lizard-like tuataras of New Zealand, and finally the lizards themselves. Snakes are a member of the lizard branch, so we will concentrate on them from now on.

From L to R: An iguana (© David Charles Photography/Shutterstock.com), a gecko (©Fedor Selivanov/Shutterstock.com) and a Komodo dragon (© Patrick Rowlands/Shutterstock.com), showing the forked tongue that helps the animal locate prey. Not to scale

Today lizards are the most abundant reptiles on earth, and are found throughout the warmer parts of the world (being "cold blooded" most reptiles find it too difficult to collect enough heat from the sun to function in colder areas). They range in size from the diminutive pygmy chameleon, not much larger than the tip of a man's finger, to the formidable Komodo dragon, which can be over 3m long and weigh 166 kg. (A relatively recent ancestor of the Komodo, known as *Megalania*, grew to some 7.6m long and weighed over 1800 kg). The lizard family (they are formally known as squamates, from a Latin word meaning "scaly", because lizards have scaly skins) can be roughly divided into two groups, based on the way they hunt: iguanas and their relatives catch food with their tongues and rely heavily on their eyes; and the "scleroglossans" or "hard tongues" (from Greek, this time). The scleroglossans, which include snakes, monitors and geckos, grasp their prey with their jaws and use their tongue and the Jacobson's organ located on their palate to both smell and taste the air. The forked tongue of the snake or monitor also gives a sense of direction: if the smell is stronger from one of the forks than the other, the animal knows the prey is in that direction. It has been accepted that these two groups reflect two simple evolutionary paths, but it now appears that the picture is more complicated than that, with different squamates converging on these two methods of catching prey.

Monitors and their relatives are known as varanids. They are a fairly diverse group, generally intelligent and social, ranging in size from the 10cm long short tailed goanna to the fearsome Komodo dragon. All have a very similar shape with a long sinuous body and neck, and a long tail. Like snakes, they use their long forked tongues to smell the air, and their jaws have some of the expandable features we see in snakes. Some of them are even venomous. Unlike snakes, however, they have four very robust limbs they can use to run very quickly, dig burrows, tear up termite mounds, climb trees and wrestle with one another.

The first lizards probably evolved some 200mya, around 10 December, while the earliest known snakes date to sometime after 130mya, almost a week later on our calendar. So from where in the widespread lizard family tree did the snakes spring? Like so many questions about the evolution of life on earth, this one has (so far) no definite answer. We know they are descended from an early four legged lizard, because of certain anatomical characteristics that the two groups share. Even today some of the older snake families, such as boas and pythons, retain tiny bone spurs located near the anus—the last remnants of their hind legs. Snake skeletons tend to be fairly delicate and don't always fossilise well. And while it seems easy to identify a snake because it has no limbs, this is not actually much help when it comes to the fossil record: a fossil could simply be missing limbs because of the way it was preserved (though there are some specialised structures in the vertebrae that allow scientists to determine if a fossil had limbs or not even if the limbs themselves aren't preserved). Or it could be another form of limbless reptile.

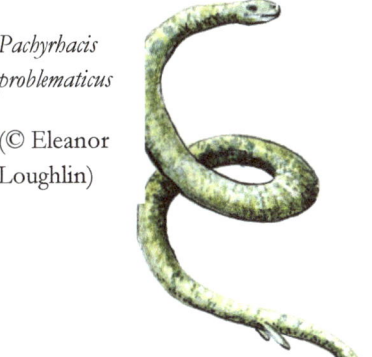

Pachyrhacis problematicus

(© Eleanor Loughlin)

A few fossil snakes from early in the evolution of the group have been found, that still retain their hind limbs. However, they seem to have raised almost as many questions as they answer. Several of them seem to have been adapted to life in the water: *Eudophis, Haasiophis* and *Pachyrhachis problematicus* (the name reflects the uncertainty about its true nature), all date to around 100-93mya. All of them have been considered aquatic, but recent research using synchrotron laminography, a very sophisticated form of X-ray, has shown that the internal structure of *Eudophis'* legs is very similar to that of modern land-living lizards. *Lapparentophis defrennei* and *Coniophis* flourished around the same time, but seem to have been terrestrial. *Najash rionegrine*, from 90my old deposits in South America, had a true pelvis as well as hind legs, and was definitely land living.

Another theory suggests that snakes evolved from ancient aquatic Varanoids, ancestors of the modern monitor lizards, such as the Gila monster and the goanna. Some of these ancient Varanoids, such as the giant Mosasaur, had a very similar jaw structure to snakes, as indeed do some modern monitors. However, some scientists feel that the similarities between snakes and Mosasaurs are the result of convergent evolution: different, unrelated animals solving the same problem in the same way, like the aardvark and anteater we mentioned in chapter five. Recent research into the DNA of living snakes and monitors also seems to show that they are not very closely related.

A monitor lizard showing the snake-like wide gape. (© Adi Soon/Shutterstock.com)

The third theory is that snakes evolved from burrowing ancestors. Streamlining yourself is certainly useful for a burrowing lifestyle, and losing your limbs is the last step in that process. Eyes are of little use underground, which is why moles are virtually blind (they also have very small limbs). The structure of the snakes' eyes is very different from that of lizards, which has led some to suggest that at some point in their evolution snakes spent so much time underground that they virtually lost their eyes. Then when they moved back towards life above ground, they re-developed their vision but in a different way. Instead of eyelids snakes' eyes are covered with a transparent scale called a brille. This may also be a remnant of a burrowing lifestyle, where the development of a transparent eye covering that allows you to see without the soil damaging your eyes would be very useful. When a snake sheds its skin the brille is shed at the same time, and you can clearly see that it is part of the scaly skin. On the other hand, the structure of the snakes' eyes are actually very similar to that of some aquatic vertebrates, which to some people reinforces the idea of an aquatic origin for snakes. In addition, a solid skull is better suited to burrowing than the flexible skull of the snake and this has led some

A shed snake skin, showing the brille over the eyes. (©Tyler Fox/Shutterstock.com)

scientists to conclude that snakes have not evolved from a burrowing ancestor.

Finally, some scientists maintain that snakes evolved from terrestrial, or land living, lizards. There is almost no fossil evidence for this theory, and some people have therefore dismissed it. However, there is no reason why a land living lizard could not have moved toward a lifestyle wriggling through soil, mud, sand or leaf litter. This would have allowed them to both avoid the attention of other terrestrial predators, and to prey on other species living in these environments.

So until more evidence is found we won't know for sure exactly what sort of creature snakes are descended from. The extinction of the dinosaurs allowed the snakes to expand greatly, and the rise of the mammals towards the end of the Cretaceous period probably helped shape their evolution as well: it appears that both venom and constriction are methods of killing prey that work particularly well on rodents, which of course are both very widespread and breed extremely prolifically. The snakes' long cylindrical shape would also have been an advantage hunting rodents, as it makes it easier to squeeze into small cracks, burrows and dense vegetation, where many rodents and other small mammals live. Of course many rodents are nocturnal, which may have contributed to the fact that snakes rely much more heavily on scent than on sight.

Today there are over 3000 species of snake around the world, found on every continent except Antarctica as well as in the sea. They range in size from the tiny blind snakes, about 10cm long, to giant pythons and anacondas that often reach 7m and can, exceptionally, be as much as 9m long. An extinct species known as Titanoboa measured some 13m long and could have been a metre in diameter. All living snakes are divided into two "infraorders": a classification smaller than the order, in this case Serpentes, but bigger than the family. The Scolecophidia includes blindsnakes and threadsnakes. They are all small, 100 cm or less in length, fossorial-or burrowing- snakes that eat insects and other invertebrates. They have other interesting features but from our point of view the most important one is that they lack the wide gape and expandable jaws that other snakes have. Unfortunately, no one is sure whether the Scolecophidia are primitive survivors from the days when all early snakes had the same anatomy, or a branch of the family that "regressed" by losing the gape.

All other snakes are included in the infraorder Alethinophidia (the literal translation of the Greek is "true snakes"), and they all have the very wide gape we are examining, though not all of them have the ability to expand their jaws. They probably first appeared towards the end of the Cretaceous period, between 90 and 65mya (22-23 December). The pipesnakes and shield tailed snakes seem to be the earliest family, and have a relatively limited gape. The boas and pythons plus all the other snakes (including the venomous species) are sometimes collectively called "macrostomata snakes", or large mouthed. This grouping includes a large majority of all the living snake species, but it can be further subdivided: considered in terms of their jaw structure the boas and pythons are transitional between the pipesnakes and the other, "advanced" species. Their jaws are fairly mobile but lack the spectacular expansion seen in many of the venomous snakes. Many of the advanced snakes have been lumped together under the name "Colubridae", and are referred to as "typical snakes". It's not a very scientific classification but it does mirror the way most of us classify snakes.

It appears that the earliest macrostomans emerged around 65mya. They were boa-like animals, some of which grew to be very large. Towards the end of the Oligocene, around 22mya, many of the early snakes died out or were greatly reduced in number, and new types appeared. Changes in climate and habitat meant there were now opportunities for new types of snake. The climate became more seasonal and grasslands spread. The dinosaurs had disappeared, and in their place new animals arose such as rodents and birds. The colubroid snakes enjoyed an "explosive radiation" and now represent about 80% of all living snakes. It is in this group that we find the most extreme examples of the specialised snake jaw structure. It also includes the most specialised of all, the venomous snakes, some of whom have modified their teeth so that they can fold back against the roof of their mouth.

It used to be assumed that the boas and pythons, with their somewhat less mobile jaws and lack of venom, were relatively primitive, while the venomous snakes were more recent and highly evolved. However, recent molecular research seems to show that the picture is not as simple as that: for instance, it seems that vipers diverged from their ancestors earlier than some other advanced groups. Just to add to the confusion, *Pachyrhachis*, the Cretaceous hind-limbed snake we discussed earlier, seems to have been a macrostomate as well. So once again we don't have enough information to determine the direction in which the snakes' unusual jaw structure evolved. Was the large gape an early characteristic that some snakes later lost? Or did it develop independently more than once? We simply don't have enough information yet to answer this question. However, for our purposes this isn't a vital issue.

The shoreline environments we mentioned above can be quite rich, so it makes sense that some reptiles evolved to exploit them. But why lose your limbs to do so? Almost all vertebrates use their limbs to move and propel themselves, to grasp and manipulate objects, or to communicate with others. Limbs are both useful and versatile. So what would be the advantage in doing without them?

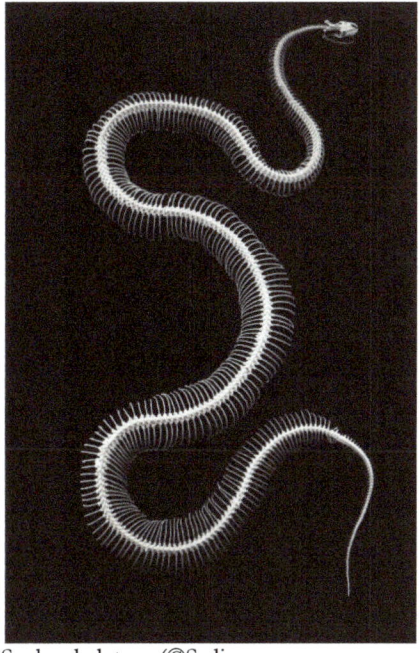

In point of fact, a number of vertebrates have lost their limbs over evolutionary time. Caecilians are legless amphibians that live in tropical areas, mostly burrowing through the soil. Slow worms, despite their name, are not worms at all but a limbless reptile related, like snakes, to lizards. No mammal is entirely without limbs, but cetaceans and other marine mammals such as dugongs are well on the way, with hind legs either absent or greatly reduced. Moles, as we mentioned above, though they retain extremely powerful clawed feet, have very short legs.

What all these animals have in common is their habitat: they spend their lives pushing through soil, fallen debris or water, and in all these cases it pays to be as streamlined as possible. Outspread limbs add to the resistance and will slow you down, besides the added risk of catching on an obstruction and possibly being damaged. Many animals that spend time in similar habitats have evolved to be long, slender and sinuous: think of ferrets, which are beautifully designed to pursue rodents through their burrows. Once you've moved far enough in that direction your body will presumably be both flexible and strong, and then losing your limbs becomes a relatively easy step. (The process appears to always happen in that order: no animal every lost its limbs first and then evolved a long sinuous body) The snakes are among the most successful limbless vertebrates, and they are beautifully adapted to this design, both inside and

Snake skeleton. (©Srdjan Draskovic/Shutterstock.com)

out. Their sleek overlapping scales protect their body while remaining flexible, and in some cases are even used in movement (see below). Internally, their skeleton has been reduced: they have no pectoral (shoulder) girdle, forelimb bones or breastbone, and only a few "primitive" species have any remnants of a pelvis or hind limbs. They can have up to 400 ribs, however. Their internal organs have adapted to the long thin shape as well: many snakes use only their right lung, which has become very elongated. Some have even lost the left lung altogether.

Snakes, with no limbs to lift or move themselves, have learned how to move in several different ways: lateral undulation involves bending the body from side to side in a wave that moves from head to tail. Where it contacts any object the snake pushes against it and uses that force to help it move forward. Sidewinding snakes move in a similar way, but lift parts of their body off the ground as they move. Some snakes move like a concertina, pulling the tail up close, moving the head forward and then repeating the movement. Large snakes such as pythons and boas actually move using the scales on their belly:

R, a sidewinder snake (© AZP Worldwide/Shutterstock.com). Its body only touches the sand at the widest points of each curve. Above, the more familiar serpentine motion(© Bogdan Ionescu/ Shutterstock.com

these are lifted up and pulled forward, then pushed down and back. They move in a straight line rather than with the undulating motion of other snakes. Accustomed to the use of limbs to move, we might be inclined to think that snakes are handicapped in this respect. In fact, they can move with surprising speed and grace, virtually dancing across sand or sliding up nearly vertical surfaces such as tree trunks. They don't actually travel very fast—the black mamba, at 6 or 7 miles per hour, is probably the fastest snake. They strike at about the same speed as a human boxer, but their reaction time is generally faster than ours so to a human the strike looks frighteningly fast.

A red squirrel using its front paws to hold food. (© Allyson Shepard Bailey)

The issues of movement aside, how much difference would the loss of their limbs make to snakes in terms of feeding? We, like all primates, are very focussed on using our hands to manipulate our environment and are at something of a loss if we can't use our hands: hence the challenge in the game of bobbing or ducking for apples. Some other animals also use their limbs to catch or manipulate their food: birds of prey use their talons, cats their grappling claws (We once had a cat who would lie on the floor and then reach over to flick bits of food out of his dish until one fell close enough to his mouth for him to eat). Dogs may use a paw to hold down a large food item while they tear off pieces with their mouth, squirrels and praying mantises hold their food in their front paws or legs. But when you look at it more closely, many creatures only use their mouths. Grazers are an obvious example, but consider a bird pecking at seeds, a frog darting its tongue at a fly, a seal catching a fish, a crocodile seizing a zebra as it crosses the river. In fact, no reptile uses its limbs to catch or hold prey. Many lizards, which in general are quite small animals, catch invertebrates with their tongues. The larger species such as the monitors seize their prey in their mouths. They may shake it, bite down on it or bang it on the ground in an attempt to either break it into smaller pieces or make it easier to ingest. To move the food down their throat they will release their jaw, sometimes tossing their heads up to help the food move. The head moves forward and to the side and the teeth take another grip. The process is repeated until the food has reached the gullet and the muscles there move it down towards the stomach. In essence, the animal walks the food down its throat. None of these reptiles makes any use of its limbs when eating. So in terms of feeding, the snakes would not have found the loss of their limbs any handicap.

So the snakes wouldn't have had any problem feeding with no limbs, but that still doesn't help explain why so many of them have ability to open their jaws wider than most other creatures. This feature seems to have been more or less built-in to the first snakes, or at least it appeared very early in their development. Why? What is the advantage? Well, it may have something to do with the underlying physics of the snake's body shape. If you streamline your body, make it longer and thinner to allow smooth movement through confined spaces, then the ratios between the different parts of the body change: the head, and therefore the size that the open mouth can attain, is relatively smaller in relation to the rest of the body than in a shorter, stouter animal. This means you need to change your feeding habits so you can take in more food in order to power the larger body. Hunting more frequently would use a lot of energy, and might be risky if your prey fought back. An alternative would be to eat more small food items—perhaps that explains the differences between the Scolecophidia and the other snakes. They may have gone down this road, concentrating on small invertebrates, and consequently either didn't develop or lost the large gape since it wasn't required for their lifestyle.

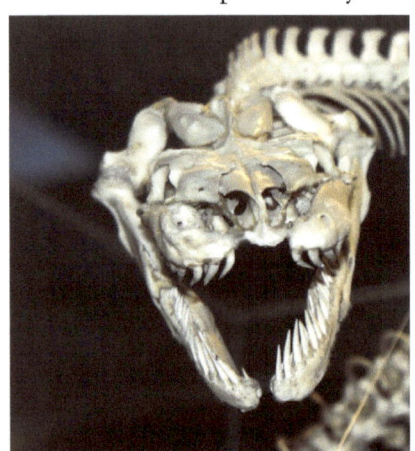

A snake's skull, showing the back curved grasping teeth and the joint at the front of the lower jaw. (© Aaron Kohr/ Shutterstock.com)

A second option would be to start catching larger prey but bite pieces out of them rather than swallowing them whole. Stoats and weasels are small but ferocious hunters and have no trouble bringing down rabbits larger than themselves, but having made the kill they have to eat their prey piece by piece. Snakes, of course, lack paws to grasp or anchor their prey, which might make it difficult to bite pieces from it—again, catching an apple in your teeth when bobbing for them is relatively easy, but it is much harder to bite a piece out of it without using your hands to hold it down. On the other hand, you could think of the snake's whole body as a manoeuvrable limb. Constrictors, obviously, are able to grasp and squeeze prey tightly enough to suffocate them. Some snakes can anchor themselves very strongly: there are cave dwelling snakes that hold onto the rocks and then reach out to catch bats on the wing. So in theory a snake could wrap its body around a food item to hold it in place while it ate it piecemeal. Unfortunately, though, snakes just don't have the right toolkit for this sort of behaviour. Their sharp, generally back-curved teeth act more like a cat's grappling claws: they can hook and hold the prey, but they aren't designed to shear through meat or bone, so they can't cut up the food they catch.

The last option is to catch larger prey and swallow it whole. Having committed themselves to their streamlined shape, the snakes presumably would be unlikely to move towards making their heads much wider than their bodies (though in fact some species do have notably triangular or diamond shaped heads). The wider you can open your jaw, the larger the prey item you can swallow. Snakes, like many reptiles, follow a pattern of eating what seems to us to be an uncomfortably large meal, and then lying torpid while it digests. Being ectothermic, or cold blooded, they don't need to continually re-fuel the way warm blooded birds or mammals do, and it is therefore much more efficient to have a single big meal rather than several small ones. Of course there could be alternative explanations: perhaps they evolved to eat large prey because the available prey was difficult to find or to catch, so they were better off catching one big meal. In any case, snakes, like many reptiles, don't have to hunt very often, and can in fact fast for weeks or months at a time.

That explains the ability to open their mouths so wide, but why did the snakes also evolve those extra "joints" in their jaw? A wide gape may allow you to take in larger prey, but unless your food is the same width as your mouth you will need to expand sideways as well. Once past the jaw a snake's throat and abdomen will stretch quite far (a snake's ribs are not joined at the front by a sternum, or breastbone, the way ours are). But the bony cage of the lower jaw presents a problem. Evolving soft tissue links at the sides and the front of the jaw allows it to expand sideways when taking in large food items—and once the food is swallowed the snake yawns, slipping everything back into place and restoring its slender silhouette.

There is another use to which snakes can put their highly mobile jaws. Remember how the monitors eat? They grip the food in their teeth, toss it loose and then grip again further along. This means there is a moment between the two bites when the food is essentially unrestrained. If the animal has not yet actually died, it might escape, or struggle in a way that could damage the predator. Or the monitor might drop its catch and lose it to another predator.

Think about the process this way: imagine you are standing at one end of a long rope and want to pull it towards you in order to reach an object tied to the other end. You can only use one hand and don't want to drop the rope at any point. So you reach forward as far as possible, grasp the rope and pull it back towards you. Once your hand is back level with your body the only option is to let go of the rope, lean forward and grasp the next section before it falls again.

Suppose, however, that you were able to use two hands, as we normally do. You reach forward with the first hand, grasp the rope and pull it back towards you. Then, still holding it, you can lean forward, grasp and pull with the second hand, and then repeat the whole cycle. At no point is the rope in danger of falling because you never fully let go of it. You can pull it towards you at a smooth, steady pace, and because you can stabilise the motion, it is less likely to become kinked or tangled or caught up in any obstruction.

This, in effect, is what the macrostomatan snakes have evolved to do. The left and right sides of both their upper and lower jaws can actually move independently. The snake can grasp the prey in its mouth, disengage the teeth on the left hand side, move them forward and bite down, then repeat the process on the right. In this way the snake inches its head and body forward engulfing the prey, which is never fully released from the snake's grip. This can of course be quite a slow process, and one theory about the development of venom in snakes suggests that, as they became able to take larger and larger prey, the danger grew that the prey animal might not be subdued or killed quickly enough, and would be able to injure the snake. Injecting it with venom ensured that it would be either dead or sufficiently paralysed to be swallowed without danger to the snake. In fact, the changes the snakes made to their jaws more or less ensured that they would have to swallow their food whole: the loosened jaw structure meant they could no longer crush their prey the way lizards do. Another side effect of the snakes' eating strategy is

69

an extremely powerful digestive system. If a snake takes too long to digest its large meals there is a real danger that the food will decay before it's fully digested, becoming toxic.

There is still one part of the process that still needs to be explained. Early snakes had established a successful lifestyle pursing small rodents and other prey in burrows and other narrow spaces. What would cause them to expand out of this niche and grow larger?

We already mentioned that the snakes, like several other groups, really became successful and widespread after the extinction of the dinosaurs. Not only did the increasing variety of mammals give them a huge new food source, the disappearance of the dinosaurs meant that there were unexploited niches ready to be filled. Nature, as everyone knows, abhors a vacuum, and when one group of animals leaves a gap in the ecological landscape, others will evolve to fill it (the formal term is "competitive release"). As we have already noticed a few times on this journey, the newcomers adapting to fill the vacated spaces will often experiment with becoming much larger. With no giant herbivorous reptiles eating the vegetation, herbivorous mammals had access to more food and could grow larger, and the carnivorous ones then had larger prey to fuel their own growth. The snakes were no different. World-wide temperatures were high enough to allow them to spread widely and grow to great size: we have already mentioned the 13m long Titanoboa. Scientists have now in fact established that only a cold blooded animal such as a reptile would be able to achieve such size with the long cylindrical body form of the snake. Warm blooded animals such as mammals would lose too much heat due to the ratio of body mass to surface area.

So the conditions were right for the snakes to become much bigger, but having evolved into that long thin shape they were faced with the problem of the size of their mouths becoming smaller in relation to the size of their body. And that brings us back to the choices an animal can make under those circumstances, as we discussed earlier.

Despite the unfilled gaps in the fossil history of snakes, we can understand the basic mechanical and physical constraints that each step in their evolution brought to bear on them. We can also see more clearly than usual how "short sighted" the evolutionary process is. We may be able to look back, but evolving life doesn't look forward. Each evolutionary change seeks to solve an immediate problem or take advantage of an immediate opportunity. But that change may cause yet another problem that has to be solved by further evolution. With the benefit of hindsight we may be able to see "better" ways to solve that first problem, but the random forces of evolution don't make plans. The snakes weren't "aiming" at the ability to swallow food larger than their own heads. The different twists and turns of their history just meant that it was the most successful way to solve the problems they faced at the time and fill the available niches.

1 Dec (310-300mya)	2 (300-290mya) Permian Hyonomus 1st hard shelled eggs	3 (290-280mya) Anapsid skulls give rise to Synapsid and Diapsid	4 (280-270mya)	5 (270-260mya)	6 (260-250mya) Triassic	7 (250-240mya)
8 (240-230mya)	9 (230-220mya)	10 (220-210mya)	11 (210-200mya)	12 (200-190mya) Jurassic Lizards	13 (190-180mya)	14 (180-170mya)
15 (170-160mya)	16 (160-150mya)	17 (150-140mya) Cretaceous Rise of mammals	18 (140-130mya)	19 (130-120mya) 1st snakes	20 (120-110mya)	21(110-100mya)
22 (100-90mya) Pachyrhachis problematicus Eudophis Haasiophis Lapparentophis defrennei Coniophis	23 (90-80mya) Najash rionegrine	24 (80-70mya)	25 (70-60mya) Palaeocene 1st macrostomatan snakes	26 (60-50mya)	27 (50-40mya) Eocene	28 (40-30mya) Oligocene
29 (30-20mya) Early snakes die out, new types emerge. Colubroid explosive radiation	30 (20-10mya) Miocene	31 (10mya-present)				

Figure 15: Snake evolution timeline

A few colourful frog species. Left to right: blue poison dart frog (©formictopus/Shutterstock.com), red eyed green tree frog (© Luis Louro/Shutterstock.com), green and black poison arrow frog (© worldswildlifewonders /Shutterstock.com), harlequin poison dart frog (© anneke/Shutterstock.com) Not to scale

CHAPTER 8: BENT OUT OF SHAPE?

Frogs and toads are among the animals most familiar to us from childhood. They appear in many stories and fairy tales, familiars of witches or kissed by princesses. To anyone of my generation Kermit the Frog is an old friend. In the real world, collecting frog spawn or tadpoles is often an early introduction to the cycles of nature. Before we go any further I need to explain that there is not, scientifically speaking, a real difference between frogs and toads. Together the group is known as the Anura, which is a Latin name that just means "without a tail". The Romans had two words, rana for frog and bufo for toad—though the latter word has only been found in one Latin author, Virgil, and may in fact refer to a field mouse! The Greeks, in contrast, only had one word, batrachus, which seems to have been used for both. French, German and English all distinguish between the two. Today, Rana is the name of a genus of frogs, and Bufo of the "true toads"—but there are anurans outside the Bufo genus that are known as toads, and even one within it that is called a frog. Basically, a toad is a frog with certain characteristics: most notable of them is a dry, sometimes warty skin and relatively short hind legs designed for walking rather than hopping. For simplicity's sake I will just say "frog" throughout this chapter.

Common frog sitting on a mass of spawned eggs (left © Birute Vijeikene/Shutterstock.com) and a Spotted-thighed poison frog carrying tadpoles on his back (© Dr. Morley Read/ Shutterstock.com)

Tree frog in a bromeliad. (©Steve Bower/Shutterstock.com)

In any case, most of us would probably consider ourselves to be quite familiar with frogs, and wouldn't find anything very remarkable about them, with the possible exception of some of the very colourful tropical species. However, when you really stop to look frogs are actually quite amazing. They reproduce in a very wide variety of ways: many lay soft gelatinous eggs that have to stay in water or at least somewhere very damp, because they don't have the waterproof shell that reptiles evolved for their eggs. Other species are direct developers – they transform into miniature adults in the egg. There are also others in which the young are carried by a parent, on or in the back, in the stomach or in the mouth. Adult frogs can be dull green or brown, or a dazzling range of colours. Many are toxic, producing poisons in their skin that range from mildly distasteful to extremely dangerous—some species are known as poison dart frogs because the Central American tribes that share their rainforest habitat use the toxin they produce to tip their hunting darts. Frogs live in most habitats around the earth though most can be found in the tropics. Most live in or near water—some have even populated the tiny pools of water that form in the cups of bromeliad flowers high in the trees of the tropical rainforest. The frogs lay their eggs there, and the young live on insects and larvae that also live in the pools, or sometimes the female frog provides them with unfertilized eggs. Despite their need for water some species even manage to live in the desert. Most frogs eat insects or other invertebrates. Some species lunge forward to seize their prey with their mouth but others sit still and shoot out their long sticky

A frog using its tongue to catch a fly (© Cathy Keifer/Shutterstock.com) and one caught in mid-leap (© Eduard Kyslynsky/Shutterstock.com). Gliding frogs have more webbing between their toes, and as they jump they spread out their feet to act like parachutes

tongue to catch their food. The tongue is attached to the front of the mouth to maximize its length (unlike our own, which is attached further back). Although we may tend to think of frogs hopping they actually move in a wide variety of ways: some just walk, others swim, climb and even glide, using the webbing between their toes as parachutes.

Many of these attributes can be found in other animals as well: birds come in many colours; some invertebrates have toxic skin or hairs, chameleons catch prey with their tongues in a similar manner to frogs. However, there is one basic characteristic of frogs that is quite unique: their rather peculiar shape. They do exhibit the standard body

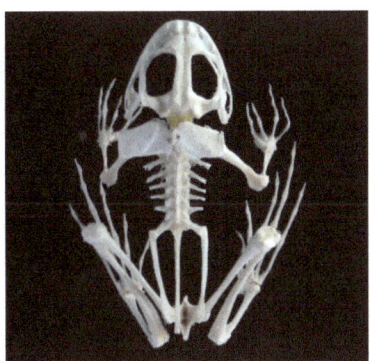

A common American bullfrog (©Allyson Shepard Bailey) and a frog skeleton (© Argonaut/Shutterstock.com). Notice the very long legs and pelvis in comparison with the short broad head and body.

plan of most land living vertebrates: a torso with four limbs at the four "corners" and a head at the front end. However, it has been adapted in a way unique to this family of animals. The head is very large in proportion to the rest of the body, as are the long, muscular hind legs. The fore legs, in contrast, are quite short. They have no visible neck, no tail and a very short stout body. When we come to look inside a frog, at the skeleton, we can see even more unusual features. Where we (and most other land living vertebrates) have two bones in our forearms, the radius and ulna; and two in our lower legs, the tibia and fibula, frogs have single bones formed by fusing the different pairs. This gives their limbs added strength to absorb the forces generated by jumping and landing. They have also lengthened two of the ankle bones (so that they actually look like the tibia and fibula in other animals), thus creating an extra joint and lever, which helps increase the power in their hind legs. Their pelvis is particularly long and has a "urostyle", a structure that is actually formed from several vertebrae fused together to make a single solid bone. This feature probably helps to transfer forces when the frog jumps. Their spine is much shorter than average (with generally nine vertebrae where we, for example, have 24). They have no ribs or else very much reduced ones. Their skulls though large are very lightly built, with far more gaps in them than we see in most other vertebrates. The long jumping hind legs and short forelegs pattern can be seen in other animals such as kangaroos and gerbils, but these animals have retained both neck and tail, as well as general proportions more recognizably like other animals. So how and why did frogs develop

Tadpoles (above © Wolfgang Staib/Shutterstock.com) and a partially metamorphosed froglet (© Dr. Morley Read/Shutterstock.com). It has developed its legs but still retains the tail

this unusual shape?

Frogs are amphibians, a group of animals whose Greek name means "both lives", because they spend part of their life in the water and the rest on land. Except in those species that develop entirely in the egg, when the young hatch out they are tadpoles, and look nothing like the parent frogs. In fact they look and behave more like fish: they have no limbs and swim using their tails for propulsion. They breathe through gills. As they mature they undergo "metamorphosis", losing their tails and gills, growing limbs, breathing air through lungs and, in most cases, moving onto the land for most of their adult lives. In a way, each tadpole re-enacts the move of the first vertebrates from water to land. Amphibians like frogs and their cousins the salamanders (and the less familiar caecilians—limbless burrowing amphibians that look rather like snakes) occupy a special place in the history of animal evolution, as they are the oldest of the vertebrate orders still alive today. They are the first descendants of those vertebrates that managed to leave the water and establish a more or less permanent life on land. In order to really understand where the frog's unusual structure came from, we need to go right back into the water and quickly trace the changes needed to make the move to land.

A goldfish (© Sarah Holmlund/Shutterstock.com), a familiar ray finned fish, and the skeleton of a related carp (© Argonaut/Shutterstock.com). Note the lack of shoulder blades and a pelvis.

Fish, the first vertebrates, developed a body structure that takes full advantage of the environment they found themselves in. Supported by the water, a fish never has to carry or move the full weight of its body. All land animals primarily use their limbs to move around (though of course some have evolved other uses such as catching prey or indeed writing books). The familiar "ray finned" fish, such as salmon or carp, don't really have limbs like that. They move their fins using muscles located in their body—the fins themselves have no muscles. These fish use their fins like the blades of an oar, to push them through the water, or to alter the direction or angle of movement. The skeleton of a fish is rather different to that of a land living animal. The pectoral girdle, which in us comprises the shoulder blades and collar bone or clavicle, in fish is actually part of the head. The pelvic girdle, or hip bones, is very small and not connected to the spine.

Panderichthys (© Eleanor Loughlin), an early tetrapodomorph

Around the early Devonian period (about 400mya, or 26 November on our calendar) some fish evolved a slightly different lifestyle and therefore a rather different body structure, including a humerus, or arm bone, that attached to the shoulder. An animal with this structure would be better able to bear its own weight as it moved around. Fossils of these fish, known as tetrapodomorphs, have been found not just in deposits from shallow lagoons and estuaries but in fresh water as well ("tetrapod", by the way, means "4-legged". These animals weren't truly 4-legged, so their name means "4-legged shaped"). Why might they have developed this new way of moving around? Well, there are several possibilities, but since the tetrapodomorphs were predators, presumably the adaptation had something to do with catching their prey. One way for a prey animal in water to escape a large predator is to flee into shallow water. The shallower the water, the less support it gives and the stronger an animal must be to lift and move itself. So presumably these fish developed their robust lobe fins in order to exploit food sources in shallow water in which their ray finned competitors wouldn't be so comfortable. With this new adaptation they could move about more strongly, with less risk of being stranded. From shallow water to land is a relatively small step to an animal that can already support its weight and move its body on its limbs. So the earliest land living vertebrates became true "tetrapods", their limbs evolving from the strengthened fins of the tetrapodomorphs. In fact, the four limbed pattern is a characteristic shared by all land vertebrates living or extinct except the few (such as whales and snakes) that have lost limbs because of the specialised lifestyle they have adopted further along in the evolutionary journey.

Of course, moving around wasn't the only challenge these new land dwellers would face. Another very fundamental problem was breathing. Fish extract oxygen from the water as it passes over their gills, but this system won't work on land. As it turns out, it is also a system that can sometimes cause problems even in water. The first colonisers of the land were plants—the earth in those early days was completely sterile, with no food source, so there was no reason for animals to leave the water. It was a different matter for plants: sunlight is their food source, and it is available in larger, undiluted amounts on land. The first fossilised plants we know of come from the Ordovician, some 488-443 mya, but the process presumably started somewhat earlier. Once they overcame the challenge of the transition from life in the water to life on land plants flourished in this huge new environment with less competition for space and little predation—at least until the invertebrates followed them. It would seem that vertebrates did not move onto the land in order to exploit this new food source. In fact they don't appear to have begun eating plants for almost 100my, until the Carboniferous period. However, all these plants may have led to another problem. When plants die and decompose, a lot of oxygen is consumed. Inevitably many of the plants would have fallen into the water when they died, and as they rotted they would have caused the oxygen levels to drop. So any animal living in the shallow water near coasts or in rivers and lakes would have had another reason to pop its head out of the water and try to make use of the free oxygen in the atmosphere. We will come back to this question of finding oxygen later in the chapter.

The mudskipper is a small tropical fish that typically inhabits estuaries and coastal areas. They have strong fleshy fins that allow them to hop around the mud flats. They also have a limited ability to breathe out of water. They absorb some oxygen through their skins, but also fill their mouths and gills with water from which they can breathe as they move about on land, returning to the water periodically to renew it—exactly as we take a breath of air and hold it before diving under the water. Some mud skippers have air bladders where they store air. Some other species of fish also breathe air, in different ways. We don't know if any tetrapodomorphs did this, but certainly in some of them there were changes to the arrangement of their nostrils. Clearly they were taking in air from above the water. Two of the main obstacles to life on land were in the process of being overcome.

Mudskipper. (© Hugh Lansdown/Shutterstock.com

By about 375mya (26 November) an animal known as *Tiktaalik roseae* had evolved. It had robust jointed fins that look, at least as far as the skeleton is concerned, very much like the limbs of a terrestrial vertebrate, but with fin rays at the end rather than digits (we distinguish between fingers and toes because in our species they are very different. But both can be called digits, and it is probably safer to use this general term when talking about other animals, particularly if we are uncertain how they used or

From top to bottom: Tiktaalik, Ichthyostega and Acanthostega (© Eleanor Loughlin) Not to scale

moved them). Other characteristics of the land dweller were the ability to move its flattened, crocodile-like head independently of its trunk, unlike fish, and a ribcage to help protect its internal organs. It probably lived mostly in shallow water.

Within 19my of *Tiktaalik* other tetrapods had evolved such as *Ichthyostega* and *Acanthostega*. They had lost the fish's scales on the back, but retained them on the belly, probably as protection when they dragged their bodies across the land. Instead of fins they had feet—but with 7 or 8 digits, rather than the five that all terrestrial vertebrates today have (or did: some, such as horses, have evolved in more recent times to lose digits). *Ichthyostega* could bear its body weight on its forelimbs, but the hind were more like paddles. Both animals probably spent most of their time in the water.

Early tetrapods such as *Ichthyostega* and *Acanthostega* looked more like a cross between a salamander and a crocodile than a frog. So really as soon as the tetrapods had established themselves they adopted the familiar body plan we see today in salamanders and many reptiles: a flattened head, long body and tail with short, generally rather splayed legs. Obviously something must have changed to lead to the frog shape.

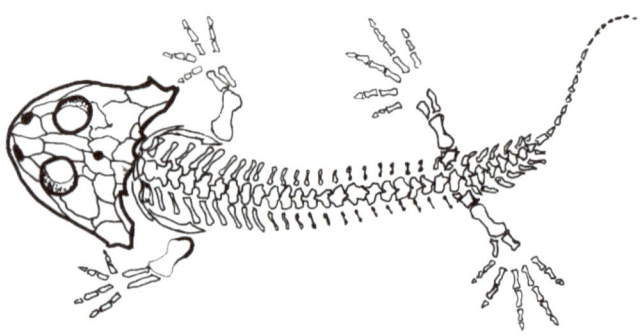

A temnospondyl skeleton. (© Eleanor Loughlin). Note the frog-like head on a salamander-like body

Ichthyostega and its relatives were succeeded by a very successful order of amphibians called the temnospondyls. They evolved during the Carboniferous, around 350mya or 27 November. Many were driven to extinction at the end of the Permian period, around 8 December. Those that survived expanded again during the Mesozoic, only to face another wave of extinctions at the Triassic/Jurassic boundary around 199mya, only 3 days later on our calendar. Once again, some survived. The most recent temnospondyls survived into the Cretaceous, perhaps around 130mya or 16 December. Like *Ichthyostega* they looked vaguely like a crocodile. As we've seen before, when a group of animals achieves dominance they often respond by increasing in size, and many of these very early amphibians were very large: *Mastodonsaurus* (the name actually means "nipple toothed lizard", though it wasn't a reptile at all) was some 6 meters long, about one-third of which was its crocodilian head. Unfortunately we don't have many fossils from this early period of amphibian development, and so we can't really tell how or indeed whether our modern amphibians are related to these early giants. However, many experts believe that the modern amphibians are descended from one branch of the temnospondyl family. Certainly the temnospondyls showed some features that are very characteristic of frogs today, such as short ribs and large holes in the roof of the mouth (these are known as interpterygoid vacuities, to use the scientific term).

Gerobatrachus (top) and *Triadobatrachus* (© Eleanor Loughlin)

Not all ancient amphibians were huge. There was a creature called *Gerobatrachus* that dates to around 290mya. It was only 7 or 8 cm long, but its discovery in 2008 caused quite a sensation (the "batrachus" part of the name comes from the Greek word for frog). It is sometimes referred to as a "frogamander" because it has the full tail and fused ankles of a salamander but the wide, light boned head of a frog. Unfortunately we don't have enough information to know if this animal was indeed an ancestor of the two modern groups. However, we do know that sometime during the Permian, maybe around 270mya, very early in December, the ancestors of frogs and salamanders split apart, so the move to "frogness" was beginning not too far from the period when *Gerobatrachus* flourished.

By 230mya an animal about 10 cm long known as *Triadobatrachus* had evolved, which had a number of features that might be called "pre-frog" adaptations. It has a fairly large skull in comparison to its body size, and like modern amphibians this consisted of fairly thin bones with a number of large openings or fenestrations. Although its front and hind legs were almost the same length, where modern frogs' have much longer hind legs, it may have been starting to jump like a frog. It probably swam using the "frog kick": that is, kicking both hind legs together rather than moving them up and down alternately. Its pelvis was quite long but lacked the urostyle we see in frogs today, and it had a tail. It also had 14 vertebrae in its backbone, where modern frogs only have between four and nine.

Throughout the Triassic period when *Triadobatrachus* was flourishing the climate was growing drier, and at the end of the period, around 200mya, some sort of catastrophe caused the extinction of many types of animals, including some of the early amphibians. But the ancestral frogs survived. A successor to *Triadoatrachus* (though we don't in fact know if this animal is a descendant) which lived around 180mya (around 10 December) was *Vieraella*, a tiny animal (30mm long) with all the obvious characteristics of modern frogs: large head with bulging eyes and long, muscular hind legs capable of impressive jumps. Its skeleton had most of the frog adaptations as well: reduced ribs, a long pelvis and urostyle, and fused bones in its legs. By the middle of the

Vieraella (© Eleanor Loughlin)

Jurassic, around 170mya, the first "true" frogs (the general name for the family is anurans, which just means "without tails") had evolved, sharing the earth with the dinosaurs as they expanded to fill the niches left vacant by the extinction event. One of the earliest, dating to between 161 and 154mya, is called *Notobatrachus*. These were somewhat larger than *Vieraella*, about 12 cm long, and display all the physical characteristics we see in frogs today. This body plan was obviously very successful: the frogs spread all over the world, occupying many different habitats and diversifying into close to 4000 different species. Today they vary in size from the tiny gold frog of Brazil, which is only 9.8mm long, to the goliath frog found in Cameroon, which can be as much as 30cm long. They can be dull shades of grey or brown, with rough warty skin, or any of a number of jewel-like colours. But all of them have retained the same unusual frog body plan that developed over 160mya.

The frog's shape may be unique in the animal world, but it has obviously also been extremely successful. We've established when modern frogs stepped onto the stage, but we still need to find out how and why. In order to do this we need to look in more detail at the skeletons of both modern frogs and their predecessors. Biologists sometimes study the skull of an animal separately from the rest of its body (technically the area below the head is called "postcranial"). Since we have identified unusual features in both the head and body of frogs, it makes sense for us to do the same.

Let's start with the head. If we go back to the early tetrapods such as *Ichthyostega* we can see that they generally had long snouted heads rather like a crocodile. By the time *Gerobatrachus* evolved the shape of the head had changed. It is short, wide and blunt nosed, with slightly bulging eyes. The skull of *Triadobatrachus* actually has more characteristics of the modern frogs than the rest of its body does. So why would the changes that would develop into that peculiar frog shape start there? One suggestion is that the large head helped to strengthen the long axis of the body, which was needed as the cervical (or neck) muscles grew larger to help ensure that the head was strongly anchored to the body. We mentioned above that *Triadobatrachus* apparently altered its pelvis to assist in swimming. It has been suggested that the proto frog's brain size increased to control this type of

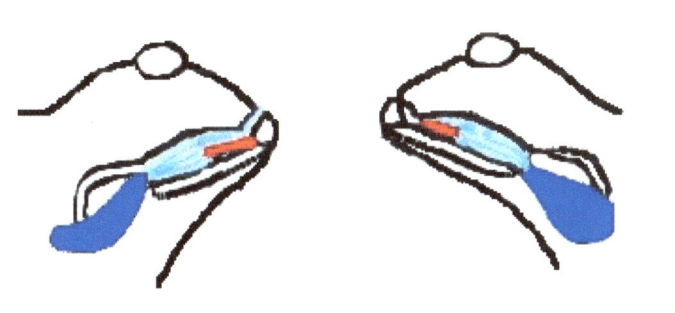

Figure 16: A very simplified diagram of frog respiration. L, the mouth is opened, allowing air (pale blue) to be drawn in. The tongue is shown in red. The lungs (dark blue) are contracted. R, the nostrils are closed and the throat is contracted. The lungs are now expanded as the air is drawn into them. The frog has gulped and swallowed air just as we gulp and swallow liquid.

movement. However, the heads of these animals weren't just getting larger in relation to their bodies. There were changes in the bones of the head as well: some fused together, while the bony palate (the roof of the mouth) was greatly reduced and had large holes in it—those are the interpterygoid vacuities we mentioned earlier. These changes are probably directly connected to the unique way that frogs breathe. When we (and most other vertebrates) breathe in, our diaphragm, a large muscle below the lungs, contracts and pulls downwards, and the rib cage expands, pulling in air so that the lungs fill. We exhale by relaxing those muscles, pushing the air out as the ribcage contracts. Frogs, on the other hand, gulp air in through the mouth and force it down the throat. In order to get as much oxygen as possible they would want to maximise the volume of the mouth chamber. That's why they have relatively large heads for the size of their bodies, and hence also those interpterygoid vacuities . Bone is rigid. Although obviously an animal's bones grow larger as it matures, it can't be moved to increase the area it encloses, except where there are joints—so when we want to take in a larger amount of air we use the joint in our jaw to open the mouth wide and yawn. By having gaps in the bones of the skull, covered with flexible skin, the frogs could raise the skin to increase the volume of air inside the mouth. Then the skin could be retracted and the eyeballs actually squeezed downwards into the mouth cavity to assist with pushing the air into the lungs.

Frog's eyes are really quite unusual: they don't just bulge out, they bulge down into the mouth cavity as well— besides using them when breathing, frogs close their eyes when they swallow, to help move the food down the throat. There is a muscle used by coelacanths today to help with swallowing. Coelacanths are sometimes referred to as "living fossils": living representatives of very ancient families that have otherwise died out. As we saw earlier, it was from fish very similar to coelacanths that the first amphibians evolved. It appears that the same muscle in the amphibian ancestor may have become, during the course of evolution, the muscle used by early amphibians to retract their eyes. That may explain how this rather peculiar mechanism developed in frogs.

We already mentioned that the climate during the Triassic was becoming drier. Oxygen levels were also falling from their very high levels at the end of the Carboniferous. Presumably the amphibians of the period had to develop mechanisms to take in more oxygen or to increase the efficiency with which they extracted oxygen from the air. Gulping air and increasing the size of the head and mouth could have helped with this. At the same time, the amphibians would have been facing increasing competition: one branch of the tetrapods had evolved into reptiles, and as the reptiles diversified they would have been trying to exploit the same food sources as the amphibians. Large eyes with colour vision are an obvious advantage when hunting prey. Having eyes near the top of your head means you can look around even when your body is submerged (think of a hippo), and if they bulge as well you get a much wider field of vision. Perhaps the frogs' bulging eyes developed to give them an edge in the local arms race.

Now let's look at the rest of the frog's body, or to be technical, the postcranial skeleton. As we mentioned at the beginning of the chapter, there are five features of the frog skeleton that are either unusual or unique to them alone:
- The hind legs are much longer than the front ones (this is not unique, of course. The same thing can be seen in gerbils, kangaroos, ostriches and even humans.)
- The radius and ulna in the front legs, and the fibula and tibia in the hind, are fused together into single bones
- They have a small number of vertebrae in their spinal column, and no tail
- They have no ribs or, in some cases, very reduced ribs, sometimes fused to the spine
- Their pelvis is very long and some of the bones that were vertebrae in ancient amphibians have fused into a single long, slender bone called the urostyle.

Let's go back again to *Triadobatrachus*. Although the reconstructions look quite frog like, when we examine its skeleton we can see the differences—but also the beginnings of the anuran adaptations. Its hind legs were slightly longer than the front, though not to the extent seen in frogs today. The radius/ulna and fibula/tibia were not fused. It had 14 vertebrae (far fewer than earlier amphibians like *Ichthyostega*, but more than modern frogs) and retained a short tail. Its ribs were greatly reduced. Its pelvis was much less exaggerated than a modern frog, but was clearly beginning to assume the same shape.

By the time *Vieraella* came along, as we have seen, it exhibited all of the specialised adaptations of modern frogs. We don't know for sure if *Triadobatrachus* was an actual ancestor of *Vieraella*, but it was certainly its predecessor, and it seems that we can assume with some safety that the development of the special features we see in *Vieraella* followed a similar path. So what did these changes in the frog skeleton mean in terms of the frog's body and the way it moved? Well, long, powerful hind legs are certainly useful for jumping. When the frogs returned to the water that their proto-amphibian ancestors had left they needed to swim in a new way, and longer, stronger hind legs would be very useful there. They might also evolve if an animal came to rely mostly on its hind legs to both move and bear the weight of the body, as both ostriches and humans do today, but clearly the frogs never did that: the design of their forelegs tells us that. If you are going to jump and then land on all fours, your limbs need to be able to absorb the shock of the landing—hence the fused bones in the frogs' legs both front and rear.

It's easier to jump with a short rigid body than a long flexible one, so that may explain why the frogs shortened their spine and dispensed with ribs. Ribs can be useful in protecting some interior organs, but they have another use in vertebrates that evolved after the amphibians. We discussed above how mammals such as ourselves use the diaphragm to draw air into our lungs and then squeeze the ribcage to breathe out. The muscles used for this are attached to the ribcage. Because frogs breathe using a very different mechanism, they don't need the ribs to anchor those muscles to push the air out and so could get rid of them without too much trouble. Without ribs the spine can be shortened without having the ribcage bump into the pelvis.

The pelvis is possibly the most unusual part of the frog's skeleton. We have already established that salamanders are the closest living relatives of frogs, but it would be hard to guess that if you simply compared their pelvic structures. The salamander's pelvis looks quite similar to that of many other vertebrates: a roughly heart shaped structure, when seen from the front, with attachments for the hind legs located about 2/3 of the way down the two sides (the roughly semi-circular depression in the pelvic bone where the legs attach is known as the acetabulum). If you look at the animal from above, the pelvis is below the spine. A frog's pelvis, on the other hand, is very long and thin. Instead of hanging vertically below the spine it lies almost parallel to it. The frog's legs are attached at the acetabulum, just as the salamander's are, but in this case it is much closer to the end of the pelvis, and if you look at the animal from above you can see that the legs join the body much further back than a salamander's do. Basically, the frog's pelvis has rotated 90° backwards.

We have already seen how scientists can study young living animals for clues about the evolution of their species—remember the little hoatzin in chapter 1? The same technique has been used with frogs, in an effort to find out more about their unusual pelvic structure. When a tadpole is developing into an adult frog, its body undergoes a number of changes. Its legs grow, its tail disappears and, remarkably, its pelvis moves from the vertical position below the spine to a backward-pointing one. This change, some scientists feel, mirrors the evolution of the early amphibians that led to frogs. Moving the pelvis backwards and elongating it so much obviously led to changes in the shape and attachments of the thigh muscles and also how they were used. It looks as if these changes show that the early frogs were starting to use their hind legs to swim and also to jump. Actually, it's probably more accurate to say that these proto-frogs were starting to hop short distances. Different species of frogs today can move in quite different ways. Some walk, bending their bodies side to side just as the first amphibians did. Some climb, some swim (most species can swim, of course, but only a few spend the majority of their lives in the water) and many can hop-- though only some are true long distance jumpers. Real jumping seems to have evolved relatively late, and was discovered independently several times by different frogs.

So why make such radical changes to your pelvis and legs? I think the process may have gone something like this: the early species of amphibians, before the frog shape evolved, probably walked rather like salamanders do today, bending their bodies from side to side into an S shape, with their legs splayed out on either side. They probably also caught their prey the same way amphibians do now, with a sudden lunge forward. Over time there would have been increasing competition for food and also more threats of predation both from other amphibians and from the expanding numbers of reptiles. Sometime before *Triadobatrachus*, an amphibian evolved with slightly longer hind legs. Perhaps it was spending more time in the water and was trying to kick more strongly, or the added power on land allowed it to move more quickly to escape a predator or to lunge at its prey. But the basic mechanics of the amphibian skeleton would pose some challenges in taking this development further.

A serval leaping (© Erik Isselée/Shutterstock.com) and an otter underwater (© Steve Estvanik/Shutterstock.com)—note the position of the hind legs

We usually think of cats stalking their prey, but sometimes they leap for it: think of a caracal catching a bird on the wing, or even a domestic house cat jumping at your bird feeder. They gather their hindquarters underneath them (effectively shortening their body, just as frogs have done) and then use their powerful hind legs to thrust them upward. In the water, look at a mammal that habitually swims, such as an otter. It swims using its powerful hind legs and webbed feet. But in general the mammal skeleton has one very noticeable difference from that of an amphibian: the legs are held directly under the body, not out to the side. When the otter kicks through the water its legs are below its body, or extended backwards in line with it. The same is true of the leaping cat: its hind legs are below the body and so all the force of the upward push is directed into that weight. And when it is airborne the hind legs extend behind the body.

A frog with its legs fully extended—compare with the serval and otter above.
(© Eduard Kyslynskyy/ Shutterstock.com)

Those early amphibians who started to invest in the power of their hind legs needed a way to concentrate the force when those legs pushed, and streamline the body during the jump (or glide, in the water). Rather than moving the attachment of legs to pelvis, to bring their legs under their body, they tipped the whole pelvis backwards and made it longer, so that the hind legs were located further back in relation to the body. When the animal pushed up into a jump, or a "frog kick" in the water, it now had the whole body in front of it. Shortening the spine and dispensing with a tail meant that when the legs were extended they were now more or less in the same position as those of a mammal: straight back in the same axis as the torso.

Jumping (or hopping) is an extremely energy efficient way to move: a hopping kangaroo uses much less energy than an antelope covering the ground at the same speed. It is also a good way to escape from a predator, since a prey animal making a sudden movement over a relatively long distance will be much harder to keep track of than one simply running away, however quickly.

We sometimes fall into the error of thinking that any animal that has stayed the same over evolutionary time is "primitive", as if it was somehow living in a developmental cul-de-sac and hasn't progressed. The real reason any species remains unchanged for a long time is because it has a very successful anatomy and way of life, and therefore hasn't had to make changes to cope with the vagaries of life that it encounters. The amphibians first appeared over 350mya, the very first vertebrates to establish a life on the land, and became the ancestors of all reptiles, birds, mammals and, of course, humans. The first modern frogs evolved about 170mya and have been essentially unchanged ever since. Their body plan may appear unusual to us, since no other animal has adopted it, but there is no denying its success. As I write (2012) frogs and all other amphibians are facing the possibility of virtual extinction in the next generation due to a fungal disease known as chytridiomycosis. Biologists all over the world are working to protect amphibian populations and develop some form of protection or cure for the disease. Let's hope they are successful, or we could see the end of this ancient and remarkable line within our own lifetimes.

10 Nov (520-510mya)	11 (510-500mya)	12 (500-490mya)	13 (490-480mya) Ordovician 1st fossilised land plants	14 (480-470mya)	15 (470-460mya)	16 (460-450mya)
17 (450-440mya) Silurian	18 (440-430mya)	19 (430-420mya)	20 (420-410mya)	21 (410-400mya)	22 (400-390mya) Devonian Tetrapodo-morphs	23 (390-380mya)
24 (380-370mya) Tiktaalik roseae	25 (370-360mya)	26 (360-350mya) Carboniferous Very high oxygen levels Ichthyostega, Acanthostega	27 (350-340mya) Temno-spondyls	28 (340-330mya)	29 (330-320mya)	30 (320-210mya)
1 Dec (310-300mya)	2 (300-290mya) Permian	3 (290-280mya) Gerobatrachus	4 (280-270mya)	5 (270-260mya) Frog/salamander split	6 (260-250mya) Triassic Drier, oxygen levels dropping	7 (250-240mya)
8 (240-230mya)	9 (230-220mya) Triadobatrachus	10 (220-210mya)	11 (210-200mya)	12 (200-190mya) Jurassic	13 (190-180mya)	14 (180-170mya) Vieraella
15 (170-160mya) 1st true modern frogs Notobatrachus	16 (160-150mya)	17 (150-140mya) Cretaceous	18 (140-130mya)	19 (130-120mya) Last temnospondyls	20 (120-110mya)	21(110-100mya)
22 (100-90mya) 1st land living plant eaters						

Figure 17: Frog evolution timeline

Colours in other branches of the animal kingdom: Blue butterfly (© Kamira/Shutterstock.com), Strawberry frog (© worldswildlifewonders/Shutterstock.com), Persian carpet flatworm (© Dan Exton/Shutterstock.com), scarlet ibis (© Allyson Shepard Bailey), green tree python (© Allyson Shepard Bailey), regal angelfish (© Vlad61/shuitterstock.com)

CHAPTER 9: COAT OF MANY COLOURS

There are, famously, no green monkeys. (Actually, the Callithrix monkey is also called the "green monkey", but its fur has only a slightly green tinge). Neither are there any pink antelope or blue wolves. In fact, considered alongside the rest of the animal kingdom, mammals are downright drab. Black, white, grey; brown, tan and russet or orange: almost all mammals have coats in some combination of these "earth tones". The only exceptions are a few whales and dolphins, which can be a bluish-grey, and a few primates such as mandrills, with their brilliant blue and red faces and even more colourful bottoms.

Almost every other branch of the animal kingdom, on the other hand, has at least some members that display, collectively, every colour of the rainbow. Think of the metallic glittering green beetles or gentian blue butterflies; yellow and purple angelfish or orange striped clownfish; black and yellow tiger salamanders or poison arrow frogs of china blue with darker blue markings; coral snakes with their red, yellow and black stripes and of course the birds: peacocks with their shimmering fan tails, birds of paradise with their extravagant plumes, rainbow lorikeets with green, yellow, red and blue plumage, even the "humble" goldfinch with its characteristic red, black and yellow markings. Even flatworms, a very ancient order of invertebrates that includes the tapeworm, boasts some species marked in beautifully undulating colourful patterns. Some animals even have colours that we are unable to see, for example in the ultraviolet.

As a member of the group, we are often accustomed to thinking of mammals as the most highly evolved of all animals on earth. Certainly they are the most recently evolved group. So it seems a little unfair that we should lack a characteristic we find so exciting in other vertebrates. Which raises the question, why should mammals lack the wide variation in colour seen in every other class of vertebrate?

It might help to turn the question around first, and ask why all the other types of animals are often so brightly coloured. Why does any animal need any colour at all? We have already seen that any feature in an animal, no matter how odd it may appear to us, must be of some use in helping it survive or reproduce. The same must be true of colour. To find the answer we need to follow up two separate but inextricably linked developments: the first is animal colouration itself: when and why animals became differently coloured, as well as examining both the physiology of their colours and how they use them. The second is the origins and function of the structure that allows colour to be perceived, the eye.

Because they are so closely linked we will have to jump back and forth between the two subjects, but I'll try to keep the discussion from becoming too confusing.

Before we start, we need to ensure that we remember that not all animals rely on vision to the same extent that we do. Humans are very visual animals, as are other primates. Quite a lot of the primate cerebral cortex (the part of the brain used for the so-called "higher functions" like learning and language) is used for deciphering signals from our eyes. People who have lost their sight require special treatment and assistance to function within our societies. The same could be said, of course, for those who have lost their hearing, but they at least can generally cross a road safely, or, in a less technologically advanced society, find their food.

An animal with eyes that lost its sight would be unlikely to survive. However, many vertebrates make much greater use of other senses than we do. Bats, according to the old saying, are blind. This is of course complete nonsense, but the insect eating varieties do rely very heavily on sound to perceive the world around them, catch their food and communicate with each other. Many animals rely heavily on other senses, some of which we don't even possess: snakes can sense the heat from their prey, dogs track by scent, fish have a "lateral line" along their sides that detects pressure differences in the water—other fish such as sharks can locate prey by the faint electrical fields they generate. We must avoid having our human prejudice in favour of vision cloud our view (I use the phrase deliberately—even in metaphor we rely on sight) of the role of colour.

Let's start at the beginning. How did animals first become coloured? Colour, unfortunately, almost never fossilises, and so we can't really know the colour of any animal that became extinct before human records, or has left no actual physical remains. Some of the chemicals that produce colour in living animals, such as haemoglobin (the red part of red blood cells) and melanin (which gives human skin its colour) have been found in fossils 150my old. Beyond that point we can make some quite well educated guesses based on established scientific laws and the evidence of living creatures that seem to have survived relatively unchanged from earlier days.

Bioluminescent jellyfish
(©William Attard
McCarthy/Shutterstock.com)

Nearly all life on earth, at least all of those forms with which we normally interact, derives its energy from the sun: plants convert it directly, some animals eat plants while others eat animals that eat plants. All life began in the sea, and of you move too far down in the water you can no longer collect the heat, light and other radiation provided by the sun. So any creature that is part of this solar powered community would be tied, at least indirectly, to visible light. (Interestingly, many of the deep sea creatures, including those that live well beyond the reach of sunlight, produce "bioluminescence", light generated from within their own bodies.)

The earliest, tiniest forms of life were microscopic and presumably didn't use colour directly, though some may have been coloured as the result of some physical or chemical properties in their bodies—which is why colonies of bacteria, for example those around the hot mud pools in Yellowstone National Park, appear as smears of brilliant colour. The light-collecting life forms, plants, generally appear green or blue because of the

Left, the Morning Glory pool in Yellowstone National Park—the colours are actually bacteria. © Galyna Andrushko/Shutterstock.com)
Right, a comb jelly—the shining colours along its limbs are the result of interference patterns from the moving cilia that line them (© John Wollwerth/Shutterstock.com)

chlorophyll inside their cells that uses the light to power the production of sugar. Some forms might have developed colour because it improved the way they absorbed or reflected heat or light. In others it is a by-product of their physical characteristics: comb jellies, planktonic creatures that can grow to be 2 meters long, produce ripples of dazzling iridescence as they move—interference patterns caused by the rippling of cilia (tiny tentacles) along their bodies.

These examples are all, so to speak, accidental: colour as the result of some physical or chemical process rather than being anything to do with the animal's appearance. And that, of course, is because as yet no animal could perceive colour: nothing had developed eyes. Obviously any living thing could be coloured if that adaptation was of use to it, whether or not anything could perceive the colour. But once animals could see, the use of colour became an important item in their inventory of tools to help them survive and reproduce. For example: ribbon worms belong to a phylum (the "nemertea") that scientists believe developed some 550 mya (very early in November). Many today are quite drab but some diurnal (meaning active during the day) species are brightly coloured. Ribbon worms do have eyes--some species have up to 250!--but most of them are only capable of distinguishing light and darkness, not colour. One popular theory is that the bright colours developed to advertise the fact that the animal tasted bad or was poisonous, in order to deter predators (a number of insects and other invertebrates do the same thing today—think of the brightly striped monarch caterpillars we visited in chapter 4). If that was the reason for the colours then presumably the predators would have to be able to distinguish

them—though patterns would still be visible even to an animal that could only see black and white. Obviously we can't know whether the ancient ancestors of the ribbon worms were as colourful as their modern descendants, but if they were it would mean that colour was being actively used very early in the history of animal life: around 7 or 8 November on our calendar. That predates nearly all vertebrates, and helps explain why so many modern groups of animals make such varied use of colour.

We've been talking about an animal's colour in a rather sweeping way, as if it were a single structure or feature, but of course it isn't. The easiest way to illustrate this is to look in the mirror. I have rather pale skin but dark, almost black, hair. You might have fair, freckled skin and auburn hair, or black skin with (almost certainly) black hair. If your dog or cat has had its fur shaved for a medical treatment you may notice that its skin colour mirrors the colour of the fur that covers it, but this is not always the case: a famous example is the polar bear, whose hair is translucent (though it appears a creamy colour) but whose skin is black. Not all animals have to passively display the colour they are born with: many change colour when they become mature, or from one season to the next, like arctic hares that are brown in the summer but white in the winter. Some animals can actually control their colour: squid have

Colourful squid
(© Melvin Lee/Shutterstock.com)

translucent skin and underneath it individual pigment cells called chromatophores. They can consciously make these cells larger or smaller, and so expand or contract individual areas of colour. The

result of this ability is that a squid's skin can display a rippling pattern of different shapes and colours, communicating its mood or intentions to others in the group or camouflaging the animal from predators. Interestingly, squid themselves are colour blind and so must respond to one anothers' patterns rather than the colours. Chameleons, of course, are famous for having a similar ability—it used to be thought that they changed to blend into their surroundings, but it now seems that the changes signal different moods, temperatures or even the health of the animal.

All of these colours are due to pigment, a chemical substance that causes a material to appear a certain colour. Much of mammal colour, including that in human skin, comes from a pigment called melanin. Another way an animal may appear coloured is from features that use "structural colour": that is, colour produced as a result of interference effects from light. Unlike pigment,

An iridescent soap bubble (© Francesco83/Shutterstock.com)

the perceived colour will vary depending on the angle at which it is viewed. Probably the most common structural colour we are familiar with is the iridescence on a soap bubble: those moving coloured patches that come and go on the surface. It isn't the soap itself that is pigmented, just the surface of the bubble reflecting the light in a way that makes us see colour. In the animal world, you can see these interference effects in a peacock's tail or some butterflies wings. Blue and green colours in many animals are most often the result of the way the reflected light is scattered. The colour related chemicals either present in their bodies, such as melanin, or ingested with their food, such as the carotenoids—which give carrots their orange colour—don't generally produce blue or green colouration. Many fish have a silvery, shiny appearance. This is caused by crystals inside the fishes' scales. In fact the fish somehow control the growth of the crystals. A school of little silver fish, moving and flashing, is very confusing for predators. This makes life much safer for each little fish, and the mirror effect of the silver scales makes each fish harder to see when it is not moving with the school.

A school of silvery ox eye scad fish (© cbpix/Shutterstock.com)

So why and when did vision evolve? Let's start with why. Image forming eyes like ours, of course, didn't spring into being fully formed. Even some single celled organisms are sensitive to light, so building on that sensitivity in more complex creatures would not be improbable. At some point an animal must have acquired some type of light sensitive cells, or photoreceptors. The Australian bearded dragon lizard today has a patch of similar cells on the top of its head, despite having two perfectly good eyes. The patch enables it to perceive, for example, when a bird of prey is descending on it from above and out of its range of vision. Scientists currently believe that the first photoreceptors evolved some 800mya (plus or minus quite a bit). The earliest ones seem to have been sensitive to ultraviolet, or UV, radiation. This sensitivity seems to have developed long enough ago to have existed in the ancestor of all living vertebrates. UV has a very short wavelength. Over time some animals shifted their perceptions to longer wavelengths, with the result that UV is now outside the range that humans can perceive, though some other animals can see it. UV, as we all know these days, is dangerous to living tissue—which is why the depletion of the ozone layer in our atmosphere, that screens UV light from the sun, is such a serious matter, and why we are constantly being exhorted to cover up and use sunscreen when we are outside. It's possible that the first photoreceptors developed to ensure the animal stayed deep enough in the water to avoid UV damage. In fact, one theory suggests that the first function of the structures that would become photoreceptors was as temperature sensors—heat and light, after all, are just different types of radiation—and where sunlight is brightest it is also most often hottest.

So almost incidentally to the ability to sense differences in temperature came the ability to perceive differences in light levels. Just being able to distinguish light from dark can be quite useful: it makes it easier to move towards the sunny upper part of the water where most of the plants will be, if you're an herbivore, and most of the prey, if you're a carnivore. It may help you to avoid collisions with other animals and objects, though other senses can do that, such as the "lateral line" of fish, which senses changes in water pressure —but this is only useful at fairly close range, while vision works at a much greater distance.

True vision, on the other hand, provides an even greater range of tools than plain light sensitivity. Locating anything by scent is a gradual process of eliminating less favourable directions, led by a marker that can move and diffuse. Sound is more directly locatable, but no plants and as far as we know few if any of these early creatures deliberately made noises, and indeed none of them have ears, though they may have been able to sense vibrations. If you can see something you can move directly towards or away from it without having to search, saving valuable time and energy. In effect, vision gives you another way to distinguish friend from foe, predator from prey, male from female.

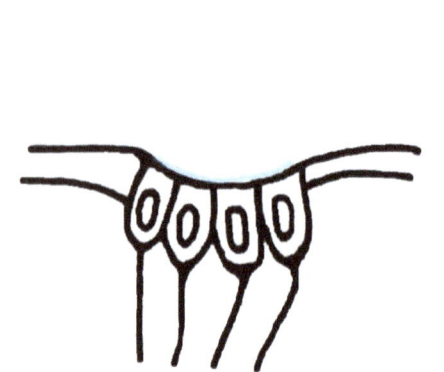 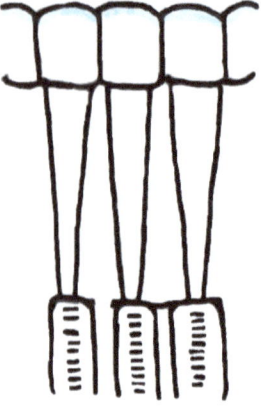 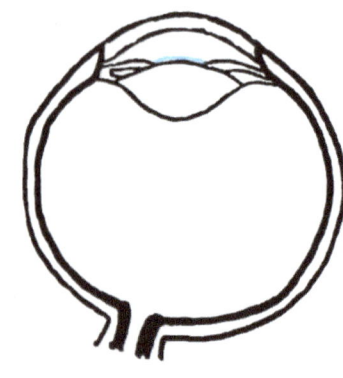

Figure 18: Simplified diagrams of three types of eye. The faint blue colour indicates where the light strikes the eye. Left, the simple "cup". The line of photoreceptors has begun to bow inwards. Centre, a compound eye. Some invertebrates may have many hundreds of these little lenses,. Right, the single lens eye. Light entering through the lens at the top is focussed on the retina at the bottom.

All vertebrate animals and a large majority of invertebrates can see (except in rare cases where they have lost that ability more recently). Eyes have been "invented" independently many times in the course of evolution, which explains why our single lens eyes are so different from the compound eyes of some insects. Our eyes, and those of many other types of animal, seem to have evolved as the original layer of photoreceptors bent inward to form a cup, which eventually became a body filled with liquid and closed by a refracting lens known as the cornea. Compound eyes may have developed in the opposite direction: the photoreceptor layer became convex, and instead of developing one large lens, the animals with compound eyes evolved an array of many lenses—a dragonfly, that flies very fast and relies on accurate vision to locate and catch its prey, has 25000 lenses in each eye. However, both types of eye are apparently initiated by the same type of gene: it seems as if there is a similar switch in many different animals marked "build an eye here", even if the resulting mechanisms are quite different. Putting the eye-building gene from a mouse into a fruit fly can actually cause the fly to develop an eye on a leg or elsewhere on its body. This seems to show that the original "build an eye" instruction goes back over 590 million years, to an ancestor shared by both fruit flies and mice. So apparently above a certain level of complexity, the ability to see becomes useful or important.

The structures and processes that make up vision are of course very complex, and we don't need to go into all the detail here. However, some understanding of the basics will be helpful. The eye acts like a camera lens, gathering light through the pupil and using the cornea and lens to focus it on the retina at the back. The reason pupils are black is nothing to do with colour—they are literally a hole leading to the inside of the eye. You get the "red eye" effect when using a flash camera because of a reflection from the blood vessels at the back of the eye bouncing back out through the pupil. That's why some cameras do a "dummy" flash first: this causes the muscles of the iris—the coloured part of the eye-- to contract so the pupils get smaller and so you don't get the red reflection. Structures in the retina translate the different wavelengths and intensity of the light into nerve impulses, which are then interpreted by the brain. The retina is covered with two types of light sensitive cells, rods and cones (named, probably unsurprisingly, after their shape). A 3rd type of photoreceptor, a cell containing a protein called melanopsin, isn't involved in image forming vision, but detects overall light levels and resets our circadian clock, the mechanism that controls when we want to sleep and wake. Melanopsin also controls how the pupil reacts to light. Rods react to the intensity of light, plain and simple, and so we, for example, use them when we see in low light without much colour. Cones react to specific wavelengths of light, and therefore to certain colours. If the light reflected from a tomato causes the cones in your eye that respond to red wavelengths to react, then you will see the tomato as red. If, however, you suffer from the visual defect known as red-green colour blindness, you will be less able to distinguish a ripe red tomato from a green unripe one. The cones use chromophores, the part of a molecule that causes a change when hit by light, and opsins, proteins that catch photons of light. Chromophores are derived from Vitamin A, which is found in certain foods including carrots. So there may be some truth to the old saying that carrots improve your vision! The chromophore accepts the photon of light and "flips", starting the process that creates an impulse to the brain. Vertebrate opsins such as our own come in five groups, one for rods and the other four for cones. Opsins combined with chromophores form "visual pigments", which absorb different wavelengths of light depending on the form of the opsin.

Probably most of us would assume that monochrome or black and white vision developed first. After all, it seems "simpler" than colour vision, and in our own experience black and white film was developed many years before colour. In that case, the first structure to evolve would be the rod, with the cone coming along later. In fact, rods are much more sensitive to light than cones are, and as far as we can tell rods actually developed from cones rather than the other way around. The first animals with true eyes only had cones. Because they had not yet developed the different opsins, however, they would still only have been able to distinguish light and dark. So the first type of vision was in fact monochrome. Fairly early on the original opsin gene had duplicated itself, so that these eyes may have already had two types of cone pigment. Even before the jawless fish had evolved into varieties with jaws, around 450mya, or 12 November, there were four cone opsin lineages.

However, it takes more than cones to see colour—your brain has to be properly wired to translate and understand the signals. In slightly more scientific terms, there needs to be some sort of comparator mechanism, something that can say "this piece of light is different from that one". In most animals this mechanism consists of neurons that operate in pairs, each one reacting to a colour that is the opposite of the one its partner reacts to. You'll have heard about opposite or complementary colours if you ever studied art: these are the colours that lie opposite one another on a traditional colour wheel. They can also be found by a visual trick: stare at a red object for 30 seconds or more. Then look at a white surface: the after image you see will be green (or more precisely, cyan, a mixture of blue and green). The comparator in your brain considers the signals from your opsins, and if they make the red "side" react more, then you see red. This pairing of colour reactions may explain why people who are colour blind generally have problems with either red and green or blue and yellow. We don't know if those animals of 450mya with their four opsins had these comparator mechanisms, but it's quite likely that they didn't really see colour.

A basic colour wheel. Colours opposite one another on the wheel are considered "complementary"

Some vertebrates also have coloured oil droplets on the inner parts of their cone receptors. These droplets act as additional filters and can improve an animal's ability to perceive colour. However, in some groups of animals, such as frogs and toads, the droplets are clear, and won't be as helpful in terms of colour vision, though they may help focus the light, and in some cases filter UV light. Other groups, such as mammals, have no oil droplets. The fact that these droplets are found in so many types of animal indicates that they probably evolved quite early—most likely before vertebrates moved onto the land, around 400mya.

We now have a general idea about the physical mechanisms that allow an animal to see colour. This leads us neatly to our second question, when did true vision start to develop? The earliest photoreceptors seem to have evolved in the first animals showing "bilateral symmetry"—

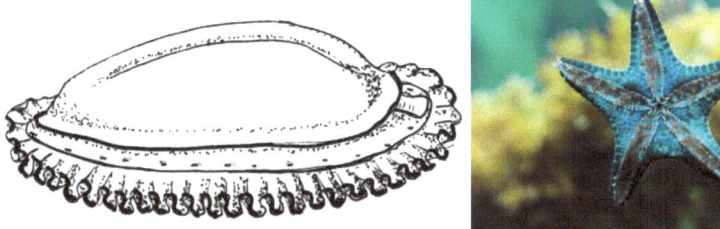

Bilateral symmetry: *Kimberella*. (© Eleanor Loughlin). Radial symmetry: starfish (©Egor Arkhipov/Shutterstock.com)

that is, with a recognisable front and back, top and bottom. The earliest known creature with bilateral symmetry is a slug-like animal called Kimberella, which seems to date to around 555mya. The ribbon worms that we mentioned above have this type of symmetry, as indeed do most living animals. In case you're wondering, it is possible to have different types of symmetry. Creatures like starfish and sea urchins have "radial symmetry", meaning that instead of having two mirror-image sides, their bodies are identical segments radiating out from the centre—rather like the 6-sided images you see in a kaleidoscope. When you start to think about it, the correlation of eyes with this type of body plan makes some sense. Once you start moving purposefully, rather than just drifting, you generally develop a front end. That end will encounter anything new first, so that's where you want to put your sense

Countershading: the manta ray's white belly tones in with the sunlit upper water (© Rich Carey/Shutterstock.com)

The baby jaguar's mottled coat mimics the dappled shadows of its surroundings (© Ronnie Howard/Shutterstock.com)

organs, the early-warning systems that tell you what is coming up. And as we discussed earlier, knowing where the sun is, whether for light or heat, and whether you want to move towards it or avoid it, will be very important. This, presumably, is how the first steps were taken on the road to vision.

We also need a date for the development of colour vision. Genetic studies seem to indicate that the five basic vertebrate visual pigments (one for rods and four for cones, remember?) had developed even before the split between vertebrates and invertebrates, some 700mya. The little oil droplets in cones apparently evolved prior to 400mya, so it's a good bet that there was some colour vision at that time. Even without more precise dating, we can accept that the ability to perceive and use colour probably developed long before any vertebrate moved onto land. It would have been firmly established long before the mammals came along.

The ability to distinguish and use colour added to the arsenal of a sighted animal. A male animal flaunting bright, conspicuous, potentially dangerous colours could send the signal "I am big, healthy and strong enough not to worry about being seen by a predator", and attract more mates. A predator that could conceal itself could be more successful ambushing prey; while a prey species that developed camouflage might avoid being eaten. We mentioned above that squid are actually colour blind but change their colour to camouflage themselves, so clearly the use of colour can be important even to an animal that can't perceive it. Either the squids' usual predators have colour vision, or the use of colour improves the matching of the squid's pattern to that of its surroundings.

Animals like squid even learned to consciously use colour, developing pigmented features within their bodies that they can control to change their colour. They send signals about their mood and intentions to others. Many fish, even if not brightly coloured, are "countershaded", with a darker back and paler belly. Seen from below, against the sky or at least the well lit upper water, they almost blend in and appear much less solid than they would if they were the same colour all over. Patterning or mottling breaks up an animal's outline, making it harder to see against its background. See a rich orange jaguar with conspicuous black rosette markings in the open and it's hard to understand why prey don't see it coming. But put the same jaguar amid the foliage in a shadow-dappled forest and it becomes all but invisible, even to our relatively sophisticated (for mammals) colour vision.

Coral reef fish are among the most colourful in the world (©Vlad61/Shutterstock.com)

Let's skip ahead to the most ancient group of vertebrates, the fish. We have already seen the uses different colouration can be put to. Colour in an animal is all about communication. They use it to advertise or to conceal. In water, where visibility decreases much more quickly with distance than it does in air, a threat or advertisement will be broadcast further with bright colours than monotones. So it makes sense that some fish would become very brightly coloured. In the feedback loop that is very common in evolution, better colour vision would encourage more use of colour—which in turn would drive improvements in vision. This ability was taken out of the water onto land by the first amphibians, who bequeathed it to their descendants, the reptiles.

Many invertebrates have colour vision as well—some butterflies have five different types of colour receptor. Mantis shrimp seem to win the prize, with up to 12 receptor types. Among vertebrates, most birds and reptiles have at least four—turtles in fact have five, which does seem a bit generous in an animal that doesn't hunt, has no need to camouflage itself and eats only vegetation. But turtles are the lone survivors of a very ancient group of reptiles, which may lend strength to the idea that the ability to differentiate many colours goes well back into vertebrate history.

Returning to the question of colour in mammals, our next step is to consider what their colour world looks like. In fact, most mammals only have two colour receptors. Some marsupials have three, while many marine mammals only have one, which is sensitive to longer wavelengths—but not the shorter blue ones, which seems odd in a largely blue environment. Among the more recently evolved mammals, only a few primates, humans among them, have three colour receptors, and that seems to be because at some point in the relatively recent past one of their two opsin genes split, so that they re-acquired a third one. (This actually happened twice independently, in our own ancestors and in the howler monkeys of South America—we'll talk more about this later). What happened in the development of mammals that caused them to lose the ancestral versatility in colour vision?

Dimetrodon, an early mammal like reptile. (© Eleanor Loughlin) The sail on its back may have been used to regulate temperature

We have already heard that mammals evolved from reptiles, but we need to look at this in more detail. The earliest "mammal like reptiles" were the pelycosaurs, that flourished around 300mya (1 December), long before the dinosaurs arose. This Permian world was becoming hotter and drier, and therefore more difficult for the amphibians that had dominated the land earlier. The climate change allowed reptiles, with their water tight eggs and skins, to become more prominent. The most familiar of the pelycosaurs is probably *Dimetrodon*, with the big sail on its back. Although *Dimetrodon* and its relatives couldn't look less like mammals, there are some features that scientists have identified as ancestral to our own, such as a lower jaw with a single bone (other reptiles have several bones in their jaw, mammals co-opted them into their ears). Most pelycosaurs were wiped out in the huge extinction at the end of the Permian period, around 251mya. But some of their descendants, the therapsids, (represented by *Lystrosaurus*, a heavily built, pig-sized creature that was found all over Gondwana) managed to survive. These started to look more like mammals, particularly in their more upright stance, with their legs held vertically under their bodies rather than sprawled to the side like those of lizards. Their jaws and teeth continued to change as well, moving more toward the familiar mammal toolkit that we see today.

A cynodont. (© Eleanor Loughlin)

The next stage in the journey is illustrated by the cynodonts (literally the name means "dog toothed). They continued the therapsid trend of vertical limbs, stronger jaws and more differentiated teeth. There is even some evidence that some of them may have had fur. Therapsids like this flourished and spread to occupy many of the niches that the dinosaurs would fill later—several times over, in fact, because the therapasids survived several more extinction events, though none was as devastating as the one at the Permian Triassic boundary. However, they were eventually pushed aside as the dinosaurs rose to dominate the land. (One theory about this takeover suggests that as the world grew hotter and drier during the Triassic, the dinosaurs showed a superior ability to retain water than the mammal like reptiles, and so were able to become more dominant). Cynodonts didn't disappear though—some of the smaller, more insignificant types survived, holding onto those niches that the generally larger dinosaurs didn't fill. One of these little cynodonts became the ancestor of all mammal lines.

Being driven into a largely nocturnal lifestyle, under the domination of much larger animals, probably contributed a great deal to the physical attributes the mammals developed. Being active at night requires good temperature control, so becoming warm blooded and covered with warm fur would have been very useful. Hunting at night requires excellent senses of smell and hearing, and the brain would have to enlarge to accommodate the enhanced senses. This, almost incidentally, could lead to other improvements in the brain that are seen in mammals today: problem solving skills, memory, physical agility etc. As it happens, a larger brain requires more energy, and so the need for more food would drive improvements in insulation and temperature regulation.

One thing that would be of little use in a nocturnal animal is good colour vision, and it was during this period that the mammals lost two of those four colour opsins. One seems to have gone after the split between reptiles and mammals, between 338 and 288mya, but prior to the split between marsupials, mammals that raise their offspring in a pouch and placental or "Eutherian" mammals, those that, like humans, keep their young inside their bodies, nourished by a placenta. These two lines divided sometime between 175 and 130mya, and it was after that point that the placental mammals lost the second cone—so some marsupials still have 3 opsins, or trichromatic vision.

By the time the dinosaurs disappeared and the mammals had the chance to expand into the empty niches, they had been forced down an evolutionary path that had changed some of their fundamental characteristics forever. Among these changes was the loss of effective colour vision. Nearly all mammals today are dichromatic, retaining just the two opsins they inherited from the first ancestral mammal. Some have even less: all cetaceans (whales and dolphins) and "pinnepeds"—animals such as seals and walrus—have basically no colour vision. Even among the dichromatic mammals there are differences in details such as the density of cones etc, so there is a wide variation how well different mammals see colour. A colour blind human, for example, still has better colour perception than a normally sighted cat.

An agouti. (© Allyson Shepard Bailey)

This brings us, at last, to the answer to our original question: why are mammals so much less colourful than other types of animals? Having lost the ability to really perceive colour at the very outset of their evolution, there was no reason for them to develop or use a wide variety of colours. What would be the point of growing vivid pink fur, when your rivals or mates would only perceive it as some variety of dull blue? The most common mammal colouration is known as "agouti", after a shy little forest dwelling rodent. The agouti gene controls the distribution of black pigment in mammal hair. Agouti type fur is generally banded grey and brown.

The tiger is one of the most colourful of living mammals, with its rich orange background and dramatic dark stripes. But compare the two photographs below: the one on the left is a "normal" tiger picture, as we see it with our trichromatic vision. But the one on the right approximates what another tiger would see: a much flatter, less vivid image.

Left: a tiger as we see it, with 3 colour vision. Right, as another tiger sees it (© Allyson Shepard Bailey) Note the reduced contrast between the orange and white fur, and how the green grass in the background appears dull brown.

Somewhere in their development, as their brains expanded and a life in the trees demanded extremely acute vision (you don't want to jump from one branch to another without being able to see where you're going!) primates re-discovered trichromatic vision. One suggestion is that an ability to see colour more accurately enabled them to tell when fruit was ripe to be eaten, as it would be easier to see, for example, red berries against the green leaves. Colour vision would also help the animals know when to eat certain leaves: many of them are toxic but may be less so when they are young and a rather different colour. So in some species a split in one of the colour opsins became a very useful new tool. It happened twice: once in an Old World primate, with the result that we and some of our primate cousins now have trichromatic vision; and again in the New World—though interestingly, there is some variation within species. Generally speaking, all males have just two opsins, while some females will have three. There is speculation that this split may actually be of use to the community, because the dichromats will be better able to notice certain colours or patterns and the trichromats others. Richard Dawkins tells the story that air crews during the second World War actually liked having a colour-blind crew member on board, because they could spot certain patterns that the others missed. And any photographer will tell you that a black and white image is actually sharper than a colour one. Howler monkeys, by some evolutionary chance, are now fully trichromatic. They live almost entirely on leaves, which as we noted before need to be chosen with care.

The fact that some primates (and of those only some of the most recently evolved) have rediscovered colour vision explains why they are the only mammals that exhibit any of the vibrant colours we find in other groups of animals. Fish and other animals that live entirely in the darkness of caves often lose not just their vision but their eyes as well. Evolution won't waste energy on building a structure that won't be used. Having been forced through the bottleneck of the dinosaur era into a nocturnal lifestyle, the mammals discarded their colour vision for the same reason. Nothing ever evolves exactly the same way twice, so those few mammals that have re-discovered colour did so by a different genetic mechanism than all previous animals. Maybe over the course of future evolution another mammal family will develop a mutation that gives them trichromatic vision; and if that trait is useful they might become more colourful themselves. But for now we will just have to accept that there are no pink antelope: because the other antelope couldn't see it.

Male mandrills have vividly coloured faces and bottoms, indicating that the species has good colour vision (© Lukich/Shutterstock.com)

13 Oct (800-790mya) 1st photo-receptors	....	23 Oct (700-690mya) Basic vertebrate visual pigments	....	7 Nov (550-540mya) Cambrian *Kimberella*	8 (540-530mya)	9 (530-520mya)
10 (520-510mya)	11 (510-500mya)	12 (500-490mya)	13 (490-480mya) Ordovician	14 (480-470mya)	15 (470-460mya)	16 (460-450mya) 4 cone opsins
17 (450-440mya) Silurian	18 (440-430mya)	19 (430-420mya)	20 (420-410mya)	21 (410-400mya)	22 (400-390mya) Devonian	23 (390-380mya)
24 (380-370mya)	25 (370-360mya)	26 (360-350mya) Carboniferous	27 (350-340mya)	28 (340-330mya) Beginning of reptile/mammal split	29 (330-320mya)	30 (320-210mya)
1 Dec (310-300mya)	2 (300-290mya) Permian Pelycosaurs	3 (290-280mya)	4 (280-270mya)	5 (270-260mya) 1st mammal opsin lost?	6 (260-250mya) Triassic	7 (250-240mya) Therapsids
8 (240-230mya)	9 (230-220mya)	10 (220-210mya)	11 (210-200mya)	12 (200-190mya) Jurassic	13 (190-180mya)	14 (180-170mya) Beginning of marsupial/placental mammal split
15 (170-160mya)	16 (160-150mya)	17 (150-140mya) Cretaceous	18 (140-130mya)	19 (130-120mya) 2nd mammal opsin lost?	20 (120-110mya)	21(110-100mya)
22 (100-90mya)	23 (90-80mya)	24 (80-70mya)	25 (70-60mya)	26 (60-50mya)	27 (50-40mya) Some primates regain colour vision	

Figure 19: Mammal colour/colour vision timeline

CONCLUSION

We began this journey in order to try to explain some features of certain animals that don't seem to make sense. With luck, you now can see that these features DO make sense, once you know more about the animals and their history. In fact, the problems never lay with the animals themselves but with our perceptions, assumptions and prejudices. There are two basic but extremely important facts people sometimes forget when thinking about these matters.

The first is: Evolution isn't over. I once worked with someone who had to break this to a friend. The friend was apparently surprised and saddened to discover that he does not, in fact, represent the pinnacle of a completed process. All around us, plants and animals continue to change just as they always have. Every extinction, whether from natural or man-made causes, leaves a gap, and the survivors will evolve to fill it. We are all, says the noted naturalist Richard Fortey, the sons and daughters of catastrophe. Even man is evolving. When humans abandoned their hunter-gatherer lifestyle for farming, roughly 10,000ya, they became smaller—the diet of the primitive farmers wasn't in fact as nutritious as what they enjoyed before. Nowadays we have made up for lost ground, and are increasing in size again. However, our artificially processed diet needs less chewing, so our jaws are becoming smaller: which is why so many people suffer pain and inconvenience from overcrowded teeth and impacted wisdom teeth. I, along with a small but growing proportion of the population, don't have and never did have any wisdom teeth. Presumably this will be true of more and more people over the coming generations, now that we have no need of heavy muscular jaws and extra robust teeth.

So what we see now is a single snapshot from a continuous film. An adaptation we feel doesn't make sense may have been developed when conditions were different and the species simply hasn't yet caught up with the new situation.

The second problem lies in the other direction, so to speak. Opponents of the Darwinian theory have always felt that one of their strongest cards is the lack of "transitional" forms. Even Darwin himself anguished over the development of the eye. Such a highly complex organ couldn't have simply appeared fully formed. On the other hand, what would any animal have done with half an eye, so to speak? How would the development of some organ that would eventually become an eye give an individual an advantage over its fellows? Critics had a field day mocking the idea of a freckle somehow becoming light sensitive and then metamorphosing into an eye. In actual fact, this, or something very similar, is exactly what happened, and eyes have independently evolved over 50 times throughout life's history. Even today, as we saw on page 85 above, the bearded dragon lizard of Australia has a light sensitive patch on the top of its head: not a "third eye", but a structure that allows it to receive advance warning of anything trying to attack it from above.

In Darwin's day, of course, the fossil record was much less fully explored than it is today. Today we not only know much more, we know just how much we don't know. Of all the uncounted billions of organisms that have lived and died throughout Earth's history, only a very, very small percentage died in suitable conditions to fossilise. Many of these, of course, would have been destroyed by natural forces, as continents shifted, mountains eroded and volcanoes erupted. So we will only ever have a few random pieces of the jigsaw. It is in fact remarkable that palaeontologists have built up as complete a picture as they have, given the dearth of specimens.

We saw when we looked at the history of whales that there were many transitional forms, and enough have been found to trace the whole journey from the land living hoofed carnivore to the fully aquatic cetaceans. So there must have been other transitions from ancient to modern species, even if we haven't found them yet. To take a hypothetical example suggested by Steve Jones in his excellent "Almost Like a Whale", butterflies flatten their wings to collect heat. Perhaps the wings developed first as skin flaps to help absorb heat, with muscle control to angle them correctly. Once these flaps reached a certain size they also became useful as wings, and their structure continued to evolve to improve their role as wings rather than as solar panels. The problem is, the transitional forms are likely to be rather less successful than those at either end of the process. A proto-butterfly with heat absorbing flaps large enough to be clumsy but not yet large enough to let the animal fly would be more vulnerable to predators than those with smaller flaps. Those that survived would be further down the road towards flight, but in the meantime fewer would actually survive and proportionately fewer would become fossilised for us to study. A lack of transitional forms doesn't mean they didn't exist, only that they weren't preserved. The mere fact that we have both fossil dinosaurs with feathers and living modern birds means there had to be transitional forms in between.

Look at the combined timeline on pages 96 and 97. Even with just the few stories we have been exploring here we can begin to see how interlinked all life on earth is. The evolutionary pressures that made early mammals small and nocturnal (and incidentally caused them to lose their colour vision) may also have contributed to the snakes' body plan. The Antarctic continental shift that allowed whales to radiate and grow so large also gave the emperor penguins an ideal place to nest—and remaining there led them to the unusual breeding strategy they use today. The Ice Ages that contributed to the extinction of the giant sloths led to the spreading grasslands with their abundant milkweed that tempted the monarch butterflies to move north.

It can sometimes be hard to remember or understand just how complex a machine is. Look into the cockpit of the space shuttle and you will be amazed and bewildered by the array of systems, controls and readouts. And of course that is only a small part of the whole structure: the entire shuttle consists of thousands of individual parts, perhaps over a million. It's hard to comprehend the technology needed to put a vehicle into orbit, allow the crew to carry out their mission and return it all to earth, but of course the shuttle didn't just appear out of nowhere. It is the culmination of years of development and invention stretching back through the early experiments with flight and rocketry, propulsion, electronics, computing, radio and television, experiments with new materials and contributions from many other disciplines as well. Not only did all of these feed into the development of the shuttle, the work has produced spin offs such as Teflon that gave us non-stick cookware and school trousers that stand up better to the punishment inflicted on them by children. Nor has the process ended there: the shuttle may have been retired, but everything that was learned during its lifetime is going into the creation of the International Space Station, and the next generation of manned space vehicles.

The space shuttle, for all its complexity, is a single human creation developed in the last generation or so. Now scale it all up to consider the entire 4 billion year history of the earth and all the life it supports. No one knows how many millions of species exist today, and they represent a very small proportion of all the species that have ever existed. Every single one of those species has been shaped and affected not only by all the other species around it, and all those that have gone before, but by the changes to the earth itself: changes in gravity and magnetism, in solar radiation and meteor activity, in geology and geography caused by tectonic activity, in climate and weather, in water supply and vegetation, in winds, tides and currents...as a machine the earth makes the space shuttle look like a paper airplane.

On our calendar of life, modern humans have only existed for the last few hours of 31 December. We have only been studying the history of the world around us for a few thousand years. Only in the last century or two has our science progressed enough for us to really begin to understand the astounding processes that made us and our world. Unfortunately, as our science progressed, so did an almost universal belief in the superiority of man. Until very recently almost all representations of the development of life on earth showed a linear progression with humans as the ultimate end product, the top of the tree. Even when we began to accept the idea that we have a relationship to and descent from other animals we named our own particular family "Primates", the "first ones", the leaders. Recently, however, attitudes have begun to change. Slowly we are moving away from this position of self centred arrogance. We are realising that we represent only the most recent development in our line and that this doesn't make us either special or superior.

We have also begun to see that just because we don't understand something, it doesn't mean it isn't right. Nothing, in fact, helps to strengthen our understanding of evolution more than "problems" such as those we have been exploring. Cats are superbly adapted predators. That fact is not actually much help in understanding the processes of evolution, because whether they evolved through a series of random changes, were created by an omnipotent deity, or came about by some other means, you'd expect they would end up being good at what they do. But something like a kangaroo up a tree is harder to explain. In fact, the only good explanation lies in the forces of natural selection: a gap in the food market has been exploited by some kangaroos, but they are still in a transitional stage from their traditional ground level grazing lifestyle. They have got up into the trees but haven't yet really adapted to them (except for being smaller than many other kangaroo species). Come back in a few million years and they may look more like rats or squirrels—or they may have found their own unique solution to the challenges of tree dwelling.

In the unforgettable words of comedian Dara O'Briain, "science knows it doesn't know everything... otherwise it would stop". We must keep looking for the "problems", because in them will be the answers we still need.

13 Oct (800-790mya) 1st photoreceptors	....	23 Oct (700-690mya) Basic vertebrate visual pigments	....	7 Nov (550-540mya) Cambrian *Kimberella*
13 (490-480mya) Ordovician 1st fossilised land plants	14 (480-470mya)	15 (470-460mya)	16 (460-450mya) 4 cone opsins	17 (450-440mya) Silurian
23 (390-380mya)	24 (380-370mya) *Tiktaalik roseae*	25 (370-360mya)	26 (360-350mya) Carboniferous Very high oxygen levels *Ichthyostega, Acanthostega*	27 (350-340mya) Temnospondyls
3 (290-280mya) *Gerobatrachus* Anapsid skullsbegin to give rise to Synapsid and Diapsid	4 (280-270mya)	5 (270-260mya) 1st mammal opsin lost? Frog/salamander split	6 (260-250mya) Triassic	7 (250-240mya) Therapsids
13 (190-180mya)	14 (180-170mya) Beginning of marsupial/placental mammal split *Vieraella*	15 (170-160mya) 1st true modern frogs *Notobatrachus*	16 (160-150mya)	17 (150-140mya) Cretaceous Flowering plants spread *Archaeopteryx*

23 (90-80mya) *Najash rionegrine*	24 (80-70mya) Xenarthran ancestor	25 (70-60mya) Palaeocene Extinction of dinosaurs 65mya 1st macrostomatan snakes	26 (60-50mya) 1st true Xenarthrans 1st Passerines in Australia/New Guinea Miacoidea True penguins First Raoellids

28 (40-30mya) Oligocene Split between armadillos and sloth/anteater branch Climate starting to cool	Hesperocyon (30mya) First cats (40mya), Aeluroidea (35mya) Midnight: Blubber replaces hair, 7:00am Mysticetes split from odontocetes , 8:00pm: mysticetes abandon teeth
30 (20-10mya) Miocene 20 genera of sloths all over the Americas, including giants Australia/New Guinea collide with Eurasia	Modern Lepidoptera Sabre tooth cats, *Pseudailurus* (20mya) Seals early echolocation, earliest known baleen

8 (540-530mya)	9 (530-520mya)	10 (520-510mya)	11 (510-500mya)	12 (500-490mya)
18 (440-430mya)	19 (430-420mya)	20 (420-410mya)	21 (410-400mya)	22 (400-390mya) Devonian Tetrapodomorphs
28 (340-330mya) Beginning of reptile/mammal split	29 (330-320mya)	30 (320-210mya)	1 Dec (310-300mya)	2 (300-290mya) Permian Pelycosaurs *Hyonomus* 1st hard shelled eggs
8 (240-230mya)	9 (230-220mya) *Triadobatrachus*	10 (220-210mya)	11 (210-200mya)	12 (200-190mya) Jurassic Earliest Lepidoptera Lizards
18 (140-130mya)	19 (130-120mya) 2nd mammal opsin lost? Last temnospondyls 1st snakes	20 (120-110mya) Gondwana breaking up, Australia and Antarctica still connected	21(110-100mya)	22 (100-90mya) 1st land living plant eaters *Pachyrhachis* and other "legged" snakes

27 (50-40mya) Eocene Some primates regain colour vision Passerine explosive radiation, Australia/New Guinea split from Antarctica, move North	Cat and dog ancestors split (42mya) Midnight: *Pakecetus*, 1:00am: *Ambulocetus*, 2:00am Remingtonocetidae, 4:00am *Procetus* 6:00am *Kutchicetus*, 9:00pm Basilosaurids
29 (30-20mya) Corvoidea split *Cynodesmus* (25mya), Amphicyonids *Proailurus* (25mya)	continental shifts begin to open up Antarctica *Janjucetus* Giant penguins Early snakes die out, new types emerge. Colubroid explosive radiation
31 (10mya-present) Pliocene, Pleistocene, Recent Giant sloths become extinct, small arboreal species survive New Guinea separates from Australia,	Tasmanian wolf and other predators extinct in New Guinea Danaus move from South the Central America (3mya), Repeated Ice Ages in North America until 10,000ya. Sabre tooth cats, *Pseudailurus*

Figure 20: combined timeline from all chapters.

Select bibliography and Links
Listed here are some of the more easily accessible and non-specialist books, articles and websites I used when writing this book. The full bibliography is available at www.peculiarpenguins.weebly.com. I hope this book has inspired you to read further and find out more about how life on our planet has changed and adapted. A word of warning though—science is constantly making new discoveries and developing new theories. Any book or article written more than 10 or 20 years ago may be using ideas that have since been proved wrong. Even some of the statements I've made may have been superseded by the time you read the book. And, of course, different authorities may have different theories or interpretations of the evidence to put forward. So always look for the most recent publications on the subject you want to explore, and bear in mind that the next one you read may say something quite different!

Ackery, P.R. and R. Vane-Wright (1984) Milkweed Butterflies. Cornell University Press

Alterton, David.(1994) Foxes, Wolves and wild Dogs of the World. Blandford Press

Attenborough, David (1979) Life on Earth. Collins BBC Books

(1984) The Living Planet   Collins BBC Books

(1990) The Trials of Life. Collins BBC Books

(1998) Life of Birds. BBC Books

(2002) Life of Mammals. BBC Books

(2009) Life Stories. Harper Collins

Bauchot, R. Ed. (1997) Snakes: a natural history. Sterling Publishing, New York

Beadle, Muriel (1977) The Cat: History, Biology and Behaviour. P. Harville

Beehler, B. (1989) "The birds of paradise" Scientific American vol. 275 no. 12 pp.117-123

Beltz, Ellin (2005) Frogs. Inside their remarkable world. Firefly books

Boorer, Michael.(1970) Wild Cats. Grosset and Dunlap

Carwardine, Mark (1986) Wild Cats. Scholastic Trade.

Chadwick, D. (2001) "Evolution of Whales" National Geographic vol. 200 number 5 pp. 64-78

Chattergee, S. (1997) The Rise of Birds. Johns Hopkins Press

Clack, Jennifer A. (2002) Gaining Ground: The Origin and Evolution of Tetrapods. Indiana University Press, Bloomington

Coppard, Kit.(1998) Big Cats. PRC Publishing

Dawkins, R. (2004) The Ancestor's Tale. Phoenix Books. London

Del Hoyo et al (eds).(1994) Handbook of Birds of the World vol. 2. Lynx Edicions

Feltwell J. (1986) Natural History of Butterflies. Christopher Helms Publishers Ltd.

Fiennes, Richard and Alice (1965) Natural History of the Dog. Natural History Press.

Firth, C. and B. Beehler (1998) The Birds of Paradise. Oxford University Press.

Goffart, M. (1971) Function and Form in the Sloth. Pergamon Press

Grace, Eric S. (1997) The nature of monarch butterflies: beauty takes flight. Greystone Books, Vancouver B.C.

Greene, H. (1997) Snakes: The evolution of mystery in nature. University of California Press.

Hallam, A. (ed) (1977) Encyclopaedia of Planet Earth Elsevier, Phaidon

Halpern, S (2001) Four Wings and a Prayer. Weidenfeld and Nicolson London

Hoffman, Don. (1989) Wandered: the monarch butterfly. Natural History Association of San Luis Obispo Coast.

Hofrichter, E. (2000) Amphibians. Firefly Books

Holland, Jennifer (2007) "Birds Gone Wild". National Geographic vol. 212 number 1 pp.88-97

Jones, Steve (1998) Almost Like a Whale. Doubleday

Kaplan, M with J Young (2010) David Attenborough's First Life. Collins, London.

Kemp, T.S. (2005) The Origin and Evolution of Mammals. Oxford University Press.

Kitchener, Andrew. (1991) The Natural History of the Wild Cat. Academic Press

LeMalo (1977) "The Emperor Penguin" American Scientist 65. Pp. 680-692

Macdonald, David (1992) The Velvet Claw. BBC

Mattison, C. (2002) The encyclopaedia of snakes. Casell Paperbacks, London.

Necker, Claire (1970) The Natural History of Cats. A.S. Barnes, New York,

O'Brien, Stephen J. and Warren E. Johnson (2007) "The Evolution of Cats: Genomic paw prints in the DNA of the world's wild cats have clarified the cat family tree and uncovered several remarkable migrations in their past" Scientific American Magazine vol. 284 no 7. Pp.68-75

Perrin, W., B. Würsig and J. Thewissen (eds) (2002) Encyclopaedia of Marine Mammals . Academic press

Pyle, R.M. (1999) Chasing Monarchs: migrating with the butterflies of passage. Houghton Mifflin Co. Boston

Resh, V. And R.T. Carde (2003) Encyclopaedia of insects. Academic Press.

Rose, Kenneth (2001) "The Ancestry of Whales". Science 21 September 2001: Vol. 293 no. 5538 pp. 2216-2217

Seigel, R and J.T. Collins (1993) Snakes: Ecology and Behavior. McGraw-Hill.

Seigel, R., J.T. Collins and S. Novak. Eds. (1987) Snakes: Ecology and evolutionary biolody. Macmillan publishing co.

Simpson, G.G. (1976) Penguins Past and Present, Here and There. Yale University Press

Smith, Charles Hamilton(1854) Natural history of Dogs

Sparks, J. and Tony Soper (1987) Penguins. David and Charles Newton Abbot

Stafford, P. (2000) Snakes. Natural History Museum

Stebbins, R. And N. Cohen (1995) A Natural History of Amphibians. Princeton University Press

Sterry, Paul (1995) Butterflies and Moths: a portrait of the animal world. Magna Books

Stewart, W and C Rothwell (1993) Paleobotany and the Evolution of Plants. Cambridge University Press.

Thewissen, J., et. al (2009) "From Land to Water: the Origin of Whales, gs, and Porpoises" Evo Edu Outreach 2:272–288

Urquhart, F. A. (1998) The Monarch Butterfly. William Caxton Ltd.

Vane-Wright, R. (2003) Butterflies. The Natural History Museum/Smithsonian

Williams, Tony (1995) The Penguins, Oxford University Press

www.monarchwatch.org

www.animals.nationalgeographic.com/animals/mammals/three-toed-sloth

http://www.talkorigins.org/faqs/vision.html

http://www.webexhibits.org/causesofcolor/7I.html

Zimmer, C. (1998) At the Water's Edge. Free Press

www.ingramcontent.com/pod-product-compliance
Lightning Source LLC
Chambersburg PA
CBHW051020180526

45172CB00002B/414